Thinking about Thought

*

The Structure of Life and the Meaning of Matter

*

Piero Scaruffi

Volume 2.
Life

*"Intelligence is not about knowing the answers
but about asking the questions"*

*"What we understand is not enough to understand
why we understand it"*

Scaruffi, Piero
Thinking about Thought - Life
All Rights Reserved © 2014 by Piero Scaruffi

ISBN-13: 978-1503362000
ISBN-10: 1503362000

(In the USA only one company is authorized to sell ISBNs: Bowker. And Bowker sells them at an outrageous price when other nations issue ISBNs for free. I consider this as, de facto, one of the most blatant scams in any industry. To protest against this government-sanctioned Bowker ISBN monopoly rip-off, I opted to obtain a free ISBN from Amazon CreateSpace, which will then appear as the publisher of this book, and I encourage all authors and publishers to do the same)

Printed and published in the US

For information: www.scaruffi.com

No part of this book may be reproduced or transmitted in any form or by any means, graphic, electronic or mechanical, including photocopying, recording, taping or by any information storage retrieval system, without the written permission of the author (http://www.scaruffi.com)

Contents

ECOLOGICAL REALISM: THE EMBODIED MIND 5

THE EVOLUTION OF LIFE: OF DESIGNERS AND DESIGNS 25

THE PHYSICS OF LIFE 74

ALTRUISM: FROM ENDOSYMBIOSIS TO SOCIOBIOLOGY 104

Preface

By the time you finish reading this book you will be a different person. I am not claiming that this book will change the way you think and act. I am simply referring to the fact that the cells in your body, including the neurons of your brain, are continuously changing. By the time you finish reading this book you will "literally" be a different body and a different brain. Every word that you read is having an effect on the connections between your neurons. And every breath you take is pacing the metabolism of your cells. This book is about what just happened to you.

This volume is one of four in a series titled "Thinking about Thought". See the first volume, "Brain", for the general preface.

Ecological Realism: The embodied mind

The Information Flow
The brain is in the body, and the body is in the world.

No living being can live in a vacuum. Life in a vacuum is an oxymoron. Life needs an environment. Life is transformation, and, without an environment, physical laws forbid transformation. Transformation requires energy from an external source. Life is a continuously changing equilibrium between an organism and its environment.

Even the most remote organism is connected to the rest of the biosphere via the air, which is the way it is mainly because of the metabolism of living beings (the atmosphere contains far more methane and other gases than chemical reactions alone would create).

From a biological perspective, the mind is one of the many organs that help an organism survive in the real world. Ultimately, from a biological perspective, the mind belongs to a body.

The mind conceives a world that is the outcome of the body's experience. The German biologist Jakob von Uexküll defined the "umwelt" of a species as the set of all possible stimuli from the environment that the species can perceive plus all possible actions that the species can perform in the environment. A species' reality is confined in its umwelt. Each species grasps reality differently because it lives inside a different umwelt.

Passive Cognition
In the 1960s the work of US biologist James-Jerome Gibson originated "ecological realism", the view that meaning is located in the interaction between living beings and the environment. Gibson started out with a critique of the traditional model of perception that harked back to Helmholtz, and ended with a new view of what cognition is.

The 19th-century German physicist Hermann von Helmholtz thought that perceptions are "unconscious inferences". Mind is isolated from the world and only knows what the senses deliver. The senses deliver signals, and the mind has to figure out how to interpret them. The mind uses whatever knowledge it has accrued. As proven by optical illusions, mind makes assumptions on such signals and "infers" what reality is. Perceptions are "hypotheses" on what reality just might be. But all of this inferring is largely invisible. Most of what goes on in the brain does not surface to the mind.

According to Gibson, instead, the process of perceiving is a process of "picking up" information that is available in the environment. The "information" that the organism uses originates from the interaction between the organism and its environment. Gibson believes that the sensory data coming from the environment already contain all the relationships needed to navigate the environment. No "representation" is needed by the brain. The brain's sole task is to "pick up" the information that the environment provides with each sensory experience.

The way this information is acquired is rather passive: the organism is free to move in the world, but it is the environment that feeds it information.

The way this information is "processed" is direct: there is no mediation by the mind. Action follows perception, and the two can be viewed as dual aspects of the same process.

Cognitive life is passive as in "no active effort to understand information".

The brain does not organize the sensory input or process the sense data. The brain is simply a tool to seek and extract information about the environment. What the brain truly does is recognize information. Information, for Gibson, is patterns: any pattern of environmental stimuli that repeats itself over time constitutes "information". Our brain is an organ capable of discovering "invariants" in the environmental stimuli.

Physically, that pattern of stimuli is a pattern of energy that impinges on the brain. A pattern of energy that flows through the sensory system, and that depends in a reliable way on the structure of the environment, is information. In other words, any pattern of energy that corresponds to the structure of the world carries information about the world. We use this principle, for example, every time we draw a line to represent the edge of an object.

Perception is a continuously ongoing process and consists in detecting the invariants of the environment. The function of the brain is to orient the organs of perception for seeking and extracting information from the continuous energy flow of the environment.

Thus perception and action are not separate processes. And perception cannot be separated from the environment in which the perceptive system (the organism) evolved and from the information which is present in that environment. Perception, action and the environment are tightly related.

Far from being simply a background for action, the environment is therefore viewed as a source of stimulation. Organisms move in the world using all the information that is available in it. Perceptual organs are not completely passive to the extent that they can orient themselves to pick up information, to "resonate" with the information in the environment. Ultimately, there is much more information in the world and less in the head than was traditionally assumed. And the environment does most of the work that we traditionally ascribe to the mind.

David Marr explained "how" we see. Gibson explained "what" we see, and why we see what we see.

Directed Cognition

The US psychologist Ulric Neisser refined this thesis. He defined cognition as the skill of dealing with knowledge that comes from the environment. The mind developed to cope with that knowledge. Neisser agrees with Gibson that organisms "pick up" information from the environment, but he differs from Gibson in that he argues in favor of directionality of exploration by the organism: the organism is not completely passive in the hands of the environment, but somehow it has a cognitive apparatus that directs its search for information.

Neisser wondered how the brain recognizes an object. Comparing a visual experience with all known objects is just too time-consuming. But comparing a visual experience with all known objects "that make sense" in that situation is much more feasible. Using a variant of Edward Tolman's cognitive maps, Neisser concluded that the brain probably "knows" in advance which objects are more likely to be "seen" in a certain situation. Some "anticipatory schemas" narrow down the set of objects that one "expects" to find in that certain situation. If I enter a restaurant, I will not expect to see a couch, but I will expect to see tables and chairs. This makes it much easier for my brain to recognize a table for a table.

Some kind of schemas accounts for how the organism can gather the available information from the environment. Between perception and action there exists a direct relation, one that is expressed by schemas. A schema is a blueprint for what information the organism presumes to encounter and what that information entails in the environment. The organism selects information from the environment based on its anticipatory schemas. "We can see only what we know how to look for". At every instant the organism constructs anticipations of information that make it easy to pick it up when it becomes available. Once picked up, information may, in turn, result in a change of the original schema, to direct further exploration of the environment. Perception is therefore a perennial cycle, from schemas to action (schemas direct action) to information (action picks up information) to schemas (information modifies schemas). The schema accounts for adaptive behavior while conserving the preeminence of cognitive processes.

Neisser's "cyclical" theory of perception also explains how the mind "filters" the huge amount of information that would exceed its capacity. An orienting schema of the nearby environment, a "cognitive map", guides the organism around the environment. A cognitive map contains schemas of the objects in the environment and spatial relations between the objects.

Perception is not about classifying objects in categories, creating concepts or any other sophisticated cognitive process. Perception is about using the information available in the surroundings for the purpose of directing action in it.

Perception and cognition transform the perceiver: an organism "is" the cognitive acts in which it engages.

Organisms and Environment

The US philosopher Fred Dretske agrees that information is in the environment and cognitive agents simply absorb it, thereby creating mental states. As Dretske puts it, information is what we can learn about the environment from a sensory signal.

From a biological standpoint, Daniel Dennett's "intentional stance" defines just the relationship between an organism and its environment. The organism continuously reflects its environment, as the organization of its system implicitly contains a representation of the environment. The organism refers to the environment. Intentionality defines an organism as a function of its beliefs and desires, which are products of natural selection. Intentional states are not internal states of the system, but descriptions of the relationship between the system and its environment.

The British psychologist Henry Plotkin defines knowledge itself as incorporating the environment. His focus is on the harmony established over the centuries between the organization and structure of a living being and the world it inhabits. Adaptation is the act of incorporating the outside world into the organism's structure and organization. More properly, this is "biological" knowledge. But human knowledge is simply a subset of biological knowledge.

This school of thought has been influential in reversing the traditional role between the living organism and the environment: the organism is no longer a protagonist, its free will unleashed and its creativity unlimited. The organism is a far more passive actor in the overall drama of Nature, one that has to rely upon (and whose behavior is conditioned by) the information that the environment supplies.

Ecological realism has also been influential in reshaping the profile of a cognitive system: since a cognitive system is simply an apparatus to pick up information and translate it into appropriate action, it turns out that pretty much any living thing can be considered, to some extent, as a cognitive system.

Life and cognition have lost some of their exclusive appeal, as we realized how constrained and passive they are.

Situation Theory

This conceptual revolution was, ultimately, about the meaning of life, and therefore affects Semantics.

The US mathematicians Jon Barwise and John Perry devised a "situational semantics" which reverses Frege's theory of meaning. According to the tradition founded at the end of the 19th century by the German mathematician and philosopher Gottlob Frege, meaning is located in the world of sense. On the contrary, Barwise and Perry anchor their theory of meaning to a biological fact: the world is full of meaning that living organisms can use. Meaning is not exclusive of language: it is pervasive in nature (e.g., smoke means fire). Meaning involves the informational content of situations and arises from regularities in the world. Reality is made of situations. Meaning arises out of recurring relations between situations. Barwise's unit of reasoning are situations because reality comes in situations. Situations are made of objects and spatio-temporal locations; objects have properties and stand in relations.

A living organism (a part of the world capable of perception and action) must be able to cope with the ever new situations thrown up during its course of events and to anticipate the future course of events. It must be able to pick up information about one situation from another situation. This can be realized by identifying similarities between situations, and relations between such similarities. Each organism performs this process of breaking down reality in a different way, as each organism "sees" reality in a different way, based on its ecological needs.

The type of a situation is determined by the regularities that the situation exhibits. Regularities are acquired by adaptation to the environment and define the behavior of an organism in the environment. The similarities between various situations make it possible for an organism to make sense of the world. At the same time they are understood by all members of the same species, by a whole "linguistic community".

Formally, one situation can contain information about another situation only if there is a relation that holds between situations sharing similarities with the former situation and situations sharing similarities with the latter situation. In that case the first situation "means" the second. Meaning is defined as relations that allow one situation to contain information about another situation.

Barwise emphasizes the "relational" nature of perception (e.g., perception is a relation between perceiver and perceived) and the "circumstantial" nature of information (information is information about the world). The mind, which processes that information, is strictly controlled by the environment.

Goal-driven Evolution
In his "teleo-evolutionary" theory, the Romanian philosopher Radu Bogdan summarized things this way: organisms are systems that are genetically programmed to maintain and replicate themselves, therefore they must guide themselves to their goals, therefore they need to obtain relevant information about their environment, therefore they need to be cognitive. It makes evolutionary sense that cognition should appear. But what "guides" the organism, and its cognition, is the environment. It could not be anything else. Cognitive systems are guided by the environment in

their goal-driven behavior. Cognitive systems are actually the very product of the evolutionary pressure of that guiding behavior.

Central to his thinking is the concept of "goal-directedness": natural selection presupposes goal-directedness. Goal-directedness arises from the genes themselves, which operate "goal-directedly". Organisms manage to survive and multiply in a hostile world by organizing themselves to achieve specific, limited goals in an ecological niche. To pursue their goals, organisms evolve ways to identify and track those goals. Such ways determine which knowledge is necessary. To obtain such knowledge, organisms learn to exploit pervasive and recurrent patterns of information in the world. The information tasks necessary to manipulate such information "select" the appropriate type of cognitive faculties that the organism must be capable of.

Imagine an organism that cannot recognize recurring situations: in every single moment of its life, it must improvise how to deal with the current situation. An organism that can recognize recurring situations can develop ways to best react to those types of situations. These "ways" make up its cognition. The organisms that survived are those whose cognition matched the situations that recur in their ecological niche.

The mind is not only controlled by the environment: it was created (or at least "selected") by the environment.

Cognition as Adaptation

The US psychologist Randy Gallistel believes that the nature of cognition lies in some "organizational" principles. Those are principles on how to organize a system so that the system can adjust rapidly and efficiently. In other words, "something" enables living organisms to make rapid adjustments of patterns of action in response to the environment. That "something" is the way they are internally organized.

No movement in nature is random. It always serves the purpose of "adapting" the state of the system to the external conditions. No matter how intelligent the action of a living being appears to be, that action satisfies the same general principle: adaptation.

At first sight, human action looks too complex to be reduced to such a simple scenario. Nevertheless, Gallistel believes that human behavior can be decomposed down to more and more elementary units, and those units do satisfy the general principle of action for the sake of adaptation only. The point is to explain how an action that looks like a whole can be decomposed in many coordinated lower-level actions. In Gallistel's hypothesis, the elementary units of behavior (reflex, oscillator, servomechanism, i.e. external stimulus to internal signal to muscle contraction) are "catalyzed" by units at the higher levels of the system.

Drawing from the model of a central program advanced by the US philosopher Paul-Alfred Weiss, Gallistel assumes that units are organized in a hierarchy that allows for competition and antagonism. A central program is a unit of behavior that is activated as a whole. A central program "selectively potentiates" subsets of lower-level units according to their relevance to the current goal. The principles that determine the "selective potentiation" of lower-level units are the same that govern the properties of elementary units. Each unit in the hierarchy appears to act independently, but it is held together in a consistent whole.

Situated Cognition

The US computer scientist Rodney Brooks, the originator of "situated cognition", shifted the emphasis of Artificial Intelligence to the interaction between an agent and its environment.

Brooks' "situated" agents have no knowledge. Their memory is not a locus of representation but simply the place where behavior is generated.

In Brooks' "subsumption" architecture, behavior is determined by the structure of the environment. The cognitive system has no need to represent the world, but only for how to operate in the world. There is no centralized function that coordinates the entire cognitive system, but a number of distributed decisional centers that operate in parallel, each of them performing a different task. The system does not have the explicit representation of what it is doing. It does have parallel processes that represent only their very limited goals.

The system decomposes in layers of goal-driven behavior, each layer being a network of finite-state automata, and incrementally composes its behavior through the interaction with the world.

Brooks can therefore account for the very fast response times required in the real world. In the real world there is no clear-cut difference between perception, reasoning and action.

Brooks turns the mind into one of many agents that live in the environment. The environment is the center of action, not the mind.

The environment is action, continuous action, continuously changing. Only a system of separate, autonomous control systems could possibly react and adapt to such a context.

The world contains all the information that the organism needs. Therefore there is no need to represent it in the mind. The environment acts like a memory external to the organism, from which the organism can retrieve any kind of information through perception.

"Intelligent" behavior can be partitioned into a set of asynchronous tasks (eating, walking, etc.), each endowed with a mechanism of perception and action. An artificial organism can be built incrementally by gradually adding new tasks. Behavior arises from layers of competence. Cognition is rational kinematics.

Brooks' ultimate point is that every intelligent being has a body!

Vehicles

The Italian neurophysiologist Valentino Breitenberg proposed a thought experiment which consists in mentally building progressively more complex machines, starting with the most elementary ones. At the beginning, there are only "vehicles" that respond to their environment. The first vehicle is simply made of a motor and a sensor: the speed of the motor is controlled by the sensor, motion is meant to be only forward. But in the real world this vehicle is subject to friction (where friction is the "metaphysical" sum of all forces of the environment) and therefore the trajectory will tend to deviate from the straight line. In fact, in a pond the movement would be quite complex. That's the whole point: despite the simple internal workings of these machines, they seem to be alive. We can increase little by little their circuitry, and at each step these vehicles seem to acquire not only new skills, but also a stronger personality.

The second vehicle is still fairly simple: two motors and two sensors. The sensor is designed to get excited by whatever kind of matter. It turns out that depending on the way they are wired, these vehicles react differently to the exciting matter: one runs towards it, while the other runs away from it. One is "aggressive", the other is "afraid".

And so forth. As their circuitry increases, the vehicles seem to exhibit more sophisticated feelings.

By adding simple electro-mechanical components, Breitenberg induces the machines to reason logically (via McCulloch-Pitts neurons). As the devices get more complicated, shapes are recognized; regularities are represented; properties of objects are discriminated. Hebbian associations (that get stronger as they are used) allow for concepts to emerge. Soon the machines start exhibiting learning and memory. Causation (as constant succession) and attention (as self-control over associations) finally lead to trains of thoughts.

At this point something very similar to the human mind can be said to be born, and all Breitenberg has to do is add circuitry for social and moral skills.

The leitmotiv of Breitenberg's research is that it is far easier to create machines that exhibit "cognitive" behavior simply by interacting with the environment than it is to analyze their behavior and try to deduce the internal structure that produces such cognitive behavior.

Breitenberg's ideas spawned an entire generation of robots, which their constructors appropriately tend to call "creatures".

Robots do not evolve

The British philosopher Andy Clark, too, wants to bring back the body into the reasoning brain.

We can dispose of the body and still find ways that a brain would calculate how to perform actions, but the very reason that we have bodies is that bodies make it a lot easier to perform those actions even without calculating every single movement. The fact that a body's movements are constrained by the body's structure is actually an advantage: once the brain directs a general action, there are only so many ways that the action can be carried out by the body. There is no need to calculate ways that are beyond the capabilities of the body.

Clark attacks the kind of Artificial Intelligence that wants to equip machines with logic and "problem solving" techniques (usually based on an abstract representation of the world.) This is a way to build a brain without taking into account the body; a very intelligent brain (possibly more intelligent than its creators) but pathetically out of touch with the reality of its body and its possible interactions with the environment.

Clark, instead, envisions a road to artificial intelligence via "autonomous agents", who are controllers of bodily action along the lines of Rodney Brooks' "subsumption architectures". They have simpler "brains", but their behavior is largely driven by their interaction with the environment instead of pure logic. Where logical systems take input from perception and calculate action, these agents use action as perception. Thus the distinction between perception and action fades away, as they are two sides of the same coin. And cognition becomes simply the interaction with the environment, not a system of logic. Learning occurs while we act. This, Clark reminds us, is more similar to the "quick and dirty" strategies employed by Nature.

All these theories of "situated agents" that criticize the "problem-solving" approach (the approach in which there is a "brain" planning everything the body does) tend to miss an important point. They look at biological organisms for inspiration, but they forget that biological organisms evolve. Robots do not evolve (yet). That is the very basic reason that Artificial Intelligence scientists came up with the need for a problem solver, for an entity capable of solving every possible problem on purely logical bases. Biological organisms embody their interaction with the environment: their body has been sculpted by evolution to optimize that interaction. It is hard to create a robot that can display the same degree of "fitness". What we do not have (yet) is evolving robots: robots that, given an environment, will build better and better fit robots. It is just very difficult to have a robot build another robot. At best, we have software inside a robot that evolves and fine-tunes itself. But that is not what happens in nature, where it is the hardware itself (not just the software) that evolves. The reason that Artificial Intelligence originally adopted the view of a "disembodied brain" is that a robot "is" disembodied: it is just a container for a mind. Our bodies are not mere containers of minds: our bodies have been shaped by evolution to be the natural object and subject of the mind.

When Clark and Brooks build their robots, they commit the same sin of the "problem-solving" crowd: they design the robot, instead of letting evolution design it. The problem-solving crowd builds a system that uses logic to decide behavior. Clark and Brooks use their own logic (the logic of their brains) to design a robot that (according to their logical thinking) will behave correctly in its environment. They have simply moved the problem-solving activity from the brain of the machine to the brain of the humans that design it. But nothing substantial has changed.

The Performatory Mind

The US psychologist Richard Carlson thinks that mental representations must have a "performatory" character, they must have to do with our body, they must be about performing an action in the environment.

Most cognitive skills are not conscious, or non-conscious (e.g., understanding language). Most cognitive activity is routine. Consciousness is necessary only when learning the skill. After it has been learned, it quickly becomes routine, unconscious routine. Introspection is actually difficult for experts, who often cannot explain why they do what they do. Most of our cognitive activity comes from a specific kind of learning: skill acquisition. Consciousness has to do with acquiring cognitive skills, which in turn depend on experiencing the world.

Cognition is embodied and situated: it is always about our body and/or our environment. Symbols and the mental processes that operate on them are grounded in sensory-motor activity.

There is continuity between symbolic awareness and perceptual-enactive awareness because symbolic representation is "performatory": it is useful precisely because it is about action; because symbols are grounded in action.

Autopoiesis

From a different perspective, similar conceptual changes were advanced in the 1960s by the Chilean neurobiologist Humberto Maturana when he argued that the relationship with the environment molds the configuration of a cognitive system ("Biology of Cognition", 1970). Maturana had worked with the US neurophysiologist Jerome Lettvin on the vision system of frogs, and they had reached the conclusion that the frog does not see "the world" but only what is relevant to the frog's survival, for example patterns of small moving shadows; and, by reacting to such patterns, the frog "catches" flies, its foodstuff while ignoring most of what we see ("What the Frog's Eye Tells the Frog's Brain", 1959). By analogy, Maturana concluded that we too see and know only what makes sense for our survival. We don't see and don't know what is out there, the objective reality, but only what our cognitive system creates. Each organism is defined by an organization, and that organization determines what the organism can perceive within the world. In turn, the organization of the system is only interested in what maintains itself. What we end up knowing, basically, is not the external world but the working of our nervous system, and that working, ultimately, is about our survival in the world.

The organization of the system produces what preserves the organization of the system. "Autopoiesis", or self-making, is this circular process.

"Autopoiesis" is the process by which living systems form and maintain their boundaries in the face of an ever-changing environment. It is the process by which an organism can continuously reorganize its own structure to interact with the world while remaining itself, no matter what the interaction is. Adaptation consists in regenerating the organism's structure so that its relationship to the environment remains constant.

Autopoiesis is not only self-organization, but also self-making.

Living systems are organized in closed loops. A living system is a network in which the function of each component is to create or transform other components while maintaining the circular organization of the whole. A cell exhibits autopoiesis, as does the Earth as a whole.

The product of a living system is a new organization of itself. It continually produces itself. The being and the doing are the same.

Autopoiesis is self-maintenance. Organisms use energy (mainly from light) and matter (water, carbon, nitrogen, etc) to continuously remake themselves.

Living systems are units of interaction. They only exist in an environment. They cannot be understood independently of their environment. They exhibit "exergonic" metabolism, which provides energy for the "endergonic" synthesis of polymers, i.e. for growth and replication.

In fact, the organism reorganizes based on environmental stimuli. The stimulus, therefore, can be viewed as that part of the environment that is absorbed by the structure.

The circular organization of living organisms constitutes a homeostatic system whose function is to maintain this very same circular organization. It is such circular organization that makes a living system a unit of interaction. At the same time, it is this circular organization that helps maintain the organism's identity through its interactions with the environment. Due to this circular organization, a living system is a self-referential system.

At the same time, a living system operates as an inductive system and in a predictive manner: its organization reflects regularities in the environment. Living systems are organized according to the principle: "what happened once will happen again".

Cognition is biological in the sense that the cognitive domain of an organism is defined by its interactions with the environment.

Cognition is the way in which an autopoietic system interacts with the environment (i.e., reorganizes itself). It is the result of the structural coupling with the environment that causes the continuous reorganization.

All living systems are cognitive systems. Cognition is simply the process of maintaining oneself by acting in the environment. Action and cognition cannot be separated: "all doing is knowing and all knowing is doing". Living is a process of cognition.

In summary, an autopoietic system is a network of transformation processes whose components interact to continuously regenerate the network. An autopoietic system holds constant its organization (its identity). Autopoiesis generates a structural coupling with the environment: the structure of the nervous system of an organism generates patterns of activity that are triggered by perturbations from the environment and that contribute to the continuing autopoiesis of the organism. Autopoiesis is necessary and sufficient to characterize a living system.

A living organism is defined by the fact that its organization makes it continually self-producing (autopoietic), i.e. not only autonomous but also self-referential ("the being and doing of an autopoietic system are inseparable").

Autopoiesis progressively generates more and more complex organisms and then intelligent organisms.

Multi-cellular organisms are born when two or more autopoietic units engage in an interaction that takes place more often than any of the interactions of each unit with the rest of the environment (a "structural coupling"). Inert elements then become macromolecules, and macromolecules become organic cells, and so on towards cellular organisms and intelligent beings.

A nervous system enables the living organism to expand the set of possible internal states and to expand the possible ways of structural coupling. But the nervous system is self-referential: perception is not representation of external world. Perception does not represent, it specifies the external world.

No living system exists independent of cognition. Each cognitive act is not about knowing the environment, but about reorganizing oneself in accordance with the environment. The autopoietic system knows only itself. There is no representation of the external world. There is just reorganization of the system based on the external world.

Intelligent behavior originates in extremely simple processes: the living cell is nothing special, but many living cells one next to the other become a complex system thanks to autopoiesis.

Therefore, autopoiesis works at different levels of organization, from the simple cell to intelligent agents.

Even life's origin can be easily explained: at some point in its history the Earth presented conditions that made the formation of autopoietic systems almost inevitable. The whole process of life depends not on the components of a living organism, but on its organization. Autopoiesis is about organization, not about the nature of the components.

Autopoiesis, being circular in nature and conservative in purpose, is at odds with evolution, which is linear in nature and non-conservative in purpose.

Maturana's evolution is then only a natural drift, a consequence of the conservation of autopoiesis and adaptation. There is no need for an external guiding force to direct evolution. All is needed is conservation of identity and capacity for reproduction.

For Maturana, information is a pointless concept. Communication is not transmission of information but rather coordination of behavior among living systems.

Maturana extends the term "linguistic" to any mutually generated domain of interactions (any "consensual domain"). When two or more living organisms interact recurrently, they generate a social coupling. Language emerges from such social coupling. In this view, language is "connotative" and not "denotative". Its function is to orient the organism within its cognitive domain.

The point Maturana reiterates is that cognition is a purely biological phenomenon. Organisms do not use any representational structures: their intelligent behavior is due only to the continuous change in their nervous system as induced by perception. Intelligence is action. Memory is not an abstract entity but simply the ability to recreate the behavior that best couples with a recurring situation within the environment.

Gaia and the Noosphere

The biosphere as a whole is autopoietic as it maintains itself through a careful balance of elements. Life (the sum of all living beings) can counter cosmological forces and make sure that the Earth continues to be a feasible habitat for life. Living beings use the chemicals available in the air and on the surface of the Earth and produce other chemicals that are released in the air and on the surface of the Earth. They do so at the rate that keeps the air and the surface of the Earth in balance with whatever cosmological forces operate on the Earth. To paraphrase the US biologist Lynn Margulis, life "is" the surface of the Earth.

It is not surprising that, in 1926, the Russian geologist Vladimir Vernadsky ranked living matter as the most powerful of geological forces (he even described how life opposes gravity's vertical pull by growing, running, swimming and even flying).

It is not surprising that the British biologist James Lovelock views the entire surface of the Earth, including "inanimate" matter, as a living being (which he named "Gaia").

Enaction

Following Maurice Merleau-Ponty's philosophical thought and drawing inspiration from Buddhist meditative practice, the Chilean philosopher Francisco Varela, a close associate of Maturana, argued in favor of an "enactive" approach to cognition: cognition as embodied action (or "enaction"), evolution not as optimal adaptation but as "natural drift". His stance views the human body both as matter and as experience, both as a biological entity and a phenomenological entity. Basically, Maturana's autopoiesis focused on the organism as a closed system, whereas Varela considers the organism as actively engaged with the environment. Movement is crucial for the organism to perceive, and movement implies a body, not just some kind of organization. Sensory-motor activity is as important as autopoiesis to explain living organisms. According to autopoiesis, cognition is mainly about the system representing itself to itself, whereas Varela views cognition as arising from the sensory-motor interaction with the environment. Maturana's theory is circular, Varela's is spiraling out towards the environment.

Varela believes in the emergent formation of direct experience without the need to posit the existence of a self. The mind is selfless. "Self" refers to a set of mental and bodily formations that are linked by causal coherence over time. At the same time the world is not a given, but reflects the actions in which we engage, i.e. it is "enacted" from our actions (or structural coupling).

Everything that exists is the projection of a brain.

Organisms do not adapt to a pre-given world. Organisms and environment mutually specify each other. Organisms drift naturally in the environment. Environmental regularities arise from the interaction between a living organism and its environment. The world of an organism is "enacted" by the history of its structural coupling with the environment. Perception is perceptually guided action (or sensorimotor enactment). Cognitive structures emerge from the recurrent sensorimotor activity that enables such a process. And perceptually guided action is constrained by the need to preserve the integrity of the organism (ontogeny) and its lineage (phylogeny).

Varela assigns an almost metaphysical meaning to Maturana's biological findings. Life is an elegant dance between the organism and the environment. The mind is the tune of that dance.

The Mind Is The Body

The US philosopher Mark Johnson rejects the theory (that he calls "objectivism") that meaning is an abstract relation between symbolic representation and objective reality, and that reason transcends the body.

Johnson, instead, believes that "imagination" is essential for human cognition, and imagination, in turn, arises from the human body. Imagination is taken to be both the creative quality (in the "Platonic" tradition) and the faculty that connects perception with reason (in the "Aristotelian" tradition). The human body is not just a machine that passively receives perceptions. It is an entity involved in a complex interaction with the world and with other bodies.

Human rationality is "embodied" because our reality is shaped by bodily movements. Johnson points out that, like all animals, we are bodies connected to the world. Whatever else we are, it comes from this basic fact, therefore from our bodily essence, from our "embodiment". Our mental life is a creation of this embodiment. It is only in the embodiment that one can find the meaning of our mental life.

His building blocks are Immanuel Kant's "schemas" (non-propositional structures of imagination), which are then amplified into concepts by metaphorical projection, and these concepts then structure and constrain our understanding and reasoning. Johnson claims that, ultimately, the body "is" the mind.

The Extended Phenotype

The US anthropologist Gregory Bateson once asked whether a blind man's cane could be considered part of the man.

A powerful metaphor to express the dependence of an organism on its environment, and the fact that the organism does not make sense without its environment, was introduced by the British biologist Richard Dawkins: the "extended phenotype" includes the world an organism interacts with.

The organism alone (the "phenotype") does not have biological relevance. What makes sense is an open system made of the organism and its neighbors. For example, a cobweb is still part of the spider. The control of an organism is never complete inside and null outside: there is rather a continuum of degrees of control, which allows partiality of control inside (e.g., parasites operate on the nervous system of their hosts) and an extension of control outside (as in the cobweb). To some extent the very genome of a species can be viewed as a representation of the environment inside the cell. Conversely, within the boundaries of an organism there can be more than one psychology (as in the case of schizophrenics).

The US philosopher Ruth Millikan went further claiming that, when determining the function of a biological "system", the "system" must include more than just the organism, something that extends beyond its skin. Furthermore, the system often needs the cooperation of other systems: the immune system can only operate if it is attacked by viruses. An organism is only a part of a larger biological system.

Tools (whether cobwebs or buckets or cars) are an extension of the organism which serve a specific purpose. Buckets store water. Cars help us move faster. Computers are an extension of the organism that serve the purpose of simulating a person or even an entire world. No matter how simple or how complex, those tools are an extension of our organism.

The model of the extended phenotype is consistent with a theory advanced by the US biologist Richard Lewontin. Each organism is the subject of continuous development throughout its life. And such development is driven by mutually interacting genes and environment. Genes per se cannot determine the phenotype, abilities or tendencies.

The organism is both the subject and the object of evolution. Organisms construct environments that are the conditions for their own further evolution and for the evolutions of nature itself towards new environments. Organism and environment mutually specify each other.

Sensory Exotica

The structure of the brain is probably directly related to the senses that the body has. Other animals have senses that humans don't have. The US psychologist Howard Hughes has provided detailed descriptions of several senses that allows animals to do things that humans cannot do.

The bat can avoid objects in absolute darkness at impressive speeds and even capture flying insects (sometimes employing a sort of somersault jump that requires incredible precision and coordination). The bat uses a high-frequency sonar system. We cannot hear it, but bats actually emit a (very high-frequency) call and then listen to its echo. We cannot hear it, but the call is very loud: the high energy is needed to maximize the range. The bat's sonar is a very accurate device: it can pinpoint a target with great accuracy even while the bat is traveling at high speed. This is possible because the sonar is used to paint a detailed picture of the surroundings. Hughes details the amount of information that is contained in the "call and echo" process and how the bat's brain picks up that information: the echo's delay is an indicator of the target's distance, the size of the object determines the loudness of the echo, the Doppler Effect allows the bat to calculate the speed of approach, and the target can be localized by comparing the two signals arriving at each ear, The bat literally "sees" with its ears. Therefore Hughes illustrates in detail how the auditory system of the bat's brain is organized. Its organization is in fact specialized for processing the echo. For example, most of the brain is devoted to processing signals at the frequency that yields the loudest echo. The bat's brain is a sophisticated computer for comparing the ultrasonic calls and their echoes, and then inferring the state of the world. The most spectacular feature of this system is actually that the bat can recognize its own echo, out of the thousands of calls and echoes that are emitted by a swarm of bats.

The biosonar is not exclusive to bats. Another environment that has very little light is the ocean. The bottom of the ocean is always dark. Mammals have more sophisticated brains than other species, and some mammals do live in oceans: dolphins and whales. Dolphins generate their sonar calls also through their nose, besides their larynx. What is significantly different between bats and dolphins is that some dolphin calls also serve as means of communication, and these "social" calls tend to be in the frequencies that are audible to humans. The dolphin call is also structurally more sophisticated (in terms of frequency components) than the bat call.

Migratory animals can orient themselves and navigate vast territories without any help from maps. An arctic bird (the tern) migrates from one pole to the other in what is the longest possible trip on Earth. Butterflies, salmons and whales are examples of wildly different species that are capable of accurate long-distance journeys (butterflies take more than a generation to complete the journey, i.e. those who begin the journey are not the ones that reach the destination).

Birds are equipped with a sixth sense for the Earth's magnetic field. They fly south in the fall and north in the spring. To accomplish their amazing long-distance feats, birds employ more than one technique. They are equipped with a sun compass and an internal clock (recognizing the position of the Sun is pointless if one doesn't know what time it is); and they are equipped with a celestial compass that can recognize the stars (or, better, the star that is at the center of the sky's nightly rotation). However, their compass is not a "polarity compass" (the common compass that always points north). Theirs is an "inclination compass": a compass about the inclination of the magnetic field relative to the force of gravity. This kind of compass is useful to figure out the latitude but it is useless to determine in which hemisphere you are (because it points to the nearest pole). This means that birds crossing the equator during their migratory journey must be able to switch the way they interpret their compass. Hughes speculates that evolution favored birds with an inclination compass because the ones with polarity compass got extinct during one of the many times in which the Earth's magnetic field flipped and the polarity reversed. This has happened 24 times in the past five million years. It turns out that birds have magnetoreceptors made of magnets in the nose.

Bees scout the territory for food, then return to the hive and communicate the location of food by performing a dance. The bees can observe the dance, decode the location and then reach that location. This process requires a combination of navigation and communication skills. The code hidden in the dance is actually the easy part: the way the bee dances conveys information about the distance of the location and its direction relative to the Sun. Watching the dance basically "programs" the bees to travel to that specific location. Bees know where the Sun is even when they cannot see it because their eyes can see ultraviolet sunlight. The pattern of polarized light in the sky depends on the position of the Sun, and the ultraviolet part of the spectrum carries the best information about the polarization of light. The photoreceptor of bees consists of cells that basically replicate the pattern of polarization in the sky: the better oriented the bee is relative to the Sun, the closer the match between the anatomy of its cells and the pattern of polarization, and the stronger the response that is generated by these cells. Just one 360-degree circle can tell the bee where the Sun is. Its compass is not magnetic but, in a sense, pure pattern matching and energy sensing.

Cephalopods can instead use skin color signals and can even change body shape in what could look like body art; and this is just one extreme form of camouflage.

Animals that live in water can use another source of information: electrical fields. Any living being swimming inside a body of water generates an electrical field. That electrical field can be used by other fish to detect who is swimming in the neighborhood. At the same time, some fish are capable of emitting their own electrical current. This current can be used for defense purposes but also as a sort of sonar (to navigate and detect prey). When the current is used as a weapon, it is just one of the many tools that nature provides animals to fight enemies. When the current is used for navigation, it represents a novel sense. Fish with passive electroreceptors are capable of sensing the electrical field generated by other fish. Fish with active electroreceptors are capable of producing an additional electrical field and of sensing the changes caused in it by the presence of other fish. The passive electroreceptors are ampullary receptors of the skin (called "Lorenzini ampullae") that are common to all fish. The active electroreceptors are more specialized tuberous receptors. Ampullary and tuberous receptors detect different features of the electrical field, respectively low frequency and high frequency features. The ampullary receptors tend to be localized in one area of the body, just like a radar, whereas tuberous receptors are spread all over the body because detecting high-frequency features requires more careful examination of the field. One can speculate that an analysis of the electrical field is enough for a fish to know not only that there is something nearby but also "what" that something is. Different objects cause different kinds of field and different variations in the field.

Pheromones are chemical messengers widely used in the animal kingdom to communicate all sorts of facts. Because they readily diffuse into the air, they can advertise the message to a broad population. The sophisticated social organization of insects (that are not capable of vocal communication) relies on pheromones. Pheromones are also commonly employed by mammals to influence sexual behavior.

Each of these senses exists because the animal's brain has a way to interpret the data and respond to them.

The Universe is a Message To Life

The picture painted by these researchers is almost opposite to the one painted by the logicians who worked on formalizing Logic: where the logician's program is based on the assumption that reason is an abstract manipulation of symbols, the "biological" program is based on the assumption that reason is bodily experience grounded in the environment. The two views could not be farther apart.

Implicit in the logician's project were the assumptions that meaning is based on truth and reference, that the mind is independent of the body, that reasoning is independent of the mind (logic exists in a world of its own, regardless of whether somebody uses it or not), and all minds use the same reasoning system. The biological approach puts the mind back firmly in the body, the body in the environment and meaning in the relationship between them. The reasoning system we use depends on our collective experience as a species and on our individual experience as bodies.

We are left to face the vast influence that the environment has on the development and evolution of the mental faculties of an organism, no less so than of its body.

The development of an organism, an ecosystem, or any other living entity, is due to interaction with the environment. In a different world, the same genomes would generate different beings. The universe is a message to life and to mind.

Further Reading

Barrett, Louise: BEYOND THE BRAIN - HOW BODY AND ENVIRONMENT SHAPE ANIMAL AND HUMAN MINDS (Princeton University Press, 2011)

Barwise, John & Perry, John: SITUATIONS AND ATTITUDES (MIT Press, 1983)

Bateson, Gregory: STEPS TO AN ECOLOGY OF MIND (Ballantine, 1972)

Bogdan, Radu: GROUNDS FOR COGNITION (Lawrence Erlbaum, 1994)

Breitenberg, Valentino: VEHICLES (MIT Press, 1984)

Brooks, Rodney & Steels, Luc: THE ARTIFICIAL LIFE ROUTE TO ARTIFICIAL INTELLIGENCE (Lawrence Erlbaum, 1995)

Carlson Richard: EXPERIENCED COGNITION (Lawrence Erlbaum, 1997)

Clancey, William: SITUATED COGNITION (Cambridge Univ Press, 1997)

Clark, Andy: BEING THERE (MIT Press, 1997)

Dawkins, Richard: THE EXTENDED PHENOTYPE (OUP, 1982)
Dennett, Daniel: THE INTENTIONAL STANCE (MIT Press, 1987)
Dretske, Fred: KNOWLEDGE AND THE FLOW OF INFORMATION (MIT Press, 1981)
Dretske, Fred: EXPLAINING BEHAVIOR (MIT Press, 1988)
Gallistel, C.R.: THE ORGANIZATION OF ACTION (Erlbaum, 1980)
Gibson, James Jerome: THE SENSES CONSIDERED AS PERCEPTUAL SYSTEMS (Houghton Mifflin, 1966)
Gibson, James Jerome: THE ECOLOGICAL APPROACH TO VISUAL PERCEPTION (Houghton Mifflin, 1979)
Helmholtz, Hermann: TREATISE ON PHYSIOLOGICAL OPTICS (1866)
Hughes, Howard: SENSORY EXOTICA (MIT Press, 1999)
Johnson, Mark: THE BODY IN THE MIND (Univ of Chicago Press, 1987)
Kitchener, Robert: PIAGET'S THEORY OF KNOWLEDGE (Yale University Press, 1986)
Lewontin, Richard: HUMAN DIVERSITY (W.H.Freeman, 1981)
Lovelock, James: GAIA (Oxford University Press, 1979)
Maturana, Humberto: AUTOPOIESIS AND COGNITION (Reidel, 1980)
Maturana, Humberto & Varela Francisco: THE TREE OF KNOWLEDGE (Shambhala, 1992)
Millikan, Ruth: LANGUAGE, THOUGHT AND OTHER BIOLOGICAL CATEGORIES (MIT Press, 1987)
Millikan, Ruth: WHAT IS BEHAVIOR? (MIT Press, 1991)
Neisser, Ulric: COGNITION AND REALITY (Freeman, 1975)
Neisser, Ulric: THE REMEMBERING SELF (Cambridge University Press, 1994)
Neisser, Ulric: THE PERCEIVED SELF (Cambridge Univ Press, 1994)
Plotkin, Henry: DARWIN MACHINES AND THE NATURE OF KNOWLEDGE (Harvard University Press, 1994)
Varela, Francisco, Thompson Evan & Rosch Eleanor: THE EMBODIED MIND (MIT Press, 1991)
Varela, Francisco: PRINCIPLES OF BIOLOGICAL AUTONOMY (North Holland, 1979)
Vernadsky, Vladimir: THE BIOSPHERE (1926)
VonUexküll, Jakob: UMWELT UND INNENWELT DER TIERE (1921)
Von Uexküll, Jakob: WORLDS OF ANIMALS AND WORLD OF MEN (1934)
Weiss, Paul: ANIMAL BEHAVIOR AS SYSTEM REACTION (1925)
Winograd, Terry & Flores, Fernando: UNDERSTANDING COMPUTERS AND COGNITION (Ablex, 1986)

THE EVOLUTION OF LIFE: OF DESIGNERS AND DESIGNS

Origins: What Was Life?

We often forget that brains are first and foremost alive, and no convincing evidence has been presented so far that dead brains can think. As far as we know, minds are alive. As far as we know, life comes first. If "thinking life" is a particular case of "life", then the same type of processes which are responsible for life may be responsible also for the mind. And the mystery of the mind, or at least the mystery of the principle that underlies the mind, may have been solved a century ago by the most unlikely sleuth: the British biologist Charles Darwin.

Darwin never really explained what he wanted to explain (the origin of species), but he probably discovered the "type of process" that is responsible for life. He called it "evolution", today we call it "design without a designer", "emergence", "self-organization" and so forth. What it means is that properties may appear when a system reorganizes itself due to external constraints, due to the fact that it has to live and survive in this world. This very simple principle may underlie as well the secret of thought. Darwin's theory of evolution is not about "survival of the fittest", Darwin's theory is about "design".

Life is defined by properties that occur in species as different as lions and bacteria. Mind would appear to be a property that differentiates humans from other living beings in a crucial way, but at closer inspection... animals do communicate, although they don't use our language; and animals do reason, although they don't use our logic; and animals do show emotions. What is truly unique about humans, other than the fact that we have developed more effective weapons to kill as many animals as we like?

Life As Maintenance

The British biologist Richard Dawkins gave this definition of life: living beings have to work to keep from eventually merging into their surroundings. That is the whole point of life.

There is a natural tendency towards merging seamlessly with the rest of nature. We have to work in order to maintain our identity. When we stop working, we die: then we merge with our surroundings.

A living being is characterized by different values for all fundamental quantities, whether temperature or density, than its surroundings.

The living being has to perform work in order to maintain that "differential" that is ultimately the essence and the meaning of life.

When the living being dies, the differential rapidly disappears and the dead being slowly dissolves, as all quantities (temperature, density, electricity, etc) become those of the surroundings.

Our habits of eating and drinking are merely a way of working to sustain that differential, in terms of energy and matter.

Living beings are never in equilibrium with their surroundings, unless they are dead.

On the contrary, nonliving things, that cannot defend themselves from the forces of nature, that cannot work "against" nature, are condemned to live in a state of equilibrium with their surroundings.

There is no border for a mountain or a sea: they flow seamlessly into a plain or a beach, whereas there is a clear border between an animal and the forest or the river it inhabits.

The Dynamics of Life

What "being alive" means is easily characterized, as we have plenty of specimens to study: life is about growing and reproducing. A living organism is capable of using the environment (sun, water, minerals, other living organisms) in order to change its own shape and size, and it is capable of creating offspring of a similar kind. In technical terms, life has two aspects: metabolism and replication. Metabolism is the interaction with the environment that results in growth. Replication is the copying of information that results in reproduction. Metabolism affects proteins, replication affects nucleid acids.

The statement that "life is growing and reproducing" is convenient for studying life on this planet, life as we know it. But certainly it would be confusing if we met aliens who speak and feel emotions but do not need to eat or go to the restrooms, and never change shape. They are born adults and they die adults. They do not even reproduce: they are born out of a mineral. Their cells do not contain genetic material. They do not make children. Would that still be "life"?

Also, that definition is not what folk psychology uses to recognize a living thing. What is an animal? Very few people would reply "something that grows and reproduces". Most people would answer "something that moves spontaneously". The "folk" definition is interesting, because it already implies a mind.

At the same time, the folk definition does not discriminate in a crisp manner between animate and inanimate matter. A rock can also move. True, it requires a "force" to move it. But so is the case with animals: they also require a force, although it is a chemical rather than a mechanical force. Animals eat and process their food to produce the chemical force that makes them move. The difference between the stone and the animal is the kind of force and where it comes from.

It is not easy to define "life". We know what life looks like on Earth: life is made of carbon, it uses left-handed chemicals, the way to create proteins is coded in DNA and uses four bases and RNA as an intermediary, life makes typos (mutations) when it copies itself, life reproduces, it grows, and it dies.

While there might well be billions of Earth-like planets in the universe, "finding" life outside the Earth is not trivial: how will we recognize it as "life" if it doesn't look like Earth life? What exactly are we looking for? Something that reproduces? something that grows? something that is made of the exact same material as life on Earth? We already have examples of each of these, whether software or crystals or lab chemicals, that we don't consider "alive". It seems that we will recognize as "life" only something that has all the properties that life has on Earth. There is an argument that this is very likely to have happened on those Earth-like planets. The support to this thesis comes from the fact that some traits did evolve independently more than once on Earth. We now that flying evolved at least four times (insects, birds, etc) and that eyes evolved independently in more than ten families of animals. Therefore one would be tempted to conclude that, for reasons that we still don't know, life tends to follow a deterministic curve that recreates more or less the same creatures or at least the same traits. At the same time, if that is the case, it is not clear why dinosaurs evolved only once: why didn't they evolve again after the mass extinction? The Earth has never seen such giant creatures again.

The contrary argument is that fixing nitrogen, a vital function for plants, never evolved (it is carried out in symbiosis with bacteria). Therefore life is not that deterministic, after all; and in that case we don't really know "what" evolved on other Earth-like planets.

The Laws Of Nature Revisited
How biology relates to the rest of the universe is less clear.

This universe exhibits an impressive spectrum of natural phenomena, some of which undergo spectacular mutations over macro or micro-time (long periods of time, or short periods of time). Life deserves a special status among them for the sheer quantity and quality of physical and chemical transformations that are involved. Nonetheless, ultimately life has to be just one of them.

Indirectly, it was Charles Darwin who started this train of thought, when he identified simple rules that Nature follows in determining how life proceeds over macro-time. While those "rules" greatly differ from the laws of Physics that (we think) govern the universe, they are natural laws of equal importance to the laws of electromagnetism or gravitation.

Why they differ so much from the others is a matter of debate. It could be that Darwin's laws are gross approximations of laws that, when discovered, will bear striking resemblance to the laws of Physics; or, conversely, maybe the laws of Physics are gross approximations of laws that, when discovered, will bear striking resemblance to the laws of evolution; or maybe they are just two different levels of explanation, one set of laws applying only to the micro-world, the other set applying to the macro-world.

A key aspect of life is that all living systems are made of the same fundamental constituents, molecules that are capable of catalyzing (speeding up) chemical reactions. But these molecules cannot move and cannot grow. Still, when they are combined in systems, they grow and move. New properties emerge. The first new property is the ability to self-assemble, to join other molecules and form new structures which are in turn able to self-assemble, triggering a cycle that leads to cells, tissues, organs, bodies, and possibly, societies and ecosystems.

In order to approach the subject of "life" in a scientific manner, we first need to discriminate among the various meanings of that term. What we normally call "life" is actually three separate phenomena. Precisely, in nature we observe three levels of organization: the phylogenetic level, which concerns the evolution over time of the genetic programs within individuals and species (and therefore the evolution of species); the ontogenetic level, which concerns the developmental process (or "growth") of a single multicellular organism; and the epigenetic level, which concerns the learning processes during an individual organism's lifetime (in particular, the nervous system, but also the immune system).

In other words, life occurs at three levels: organisms evolve into other organisms, each organism changes (or grows) from birth till death, and finally the behavior of each organism changes during its lifetime (the organism "learns").

There are therefore two aspects to the word "life". Because of the way life evolved and came to be what it is today, life is both reproduction and metabolism: it is both information that survives from one individual to another ("genotype"), and information about the individual ("phenotype"). When we say that "ants are alive" and "I am alive" we mean two different things, even if we use the same word. To unify those two meanings it takes a theory that explains both life as reproduction and life as growth.

Design Without a Designer
In 1859 Darwin published "The Origin Of Species". His claim was simple: all existing organisms are the descendants of simpler ancestors that lived in the distant past, and the main force driving this evolution is natural selection by the environment. This is possible because living organisms reproduce and vary (make children that are slightly different than the parents). Natural selection "selects" the fittest children, and the process continues, generation after generation, causing evolution. Through this process of evolution, organisms acquire characteristics that make them more "fit" to survive in their environment (or better "adapted" to their environment).

Darwin based his theory of evolution on some hard facts. The population of every species can potentially grow exponentially in size. Most populations don't. Resources are limited. Individuals of all species are unique, each one slightly different from the other. Such individual differences are passed on to offspring. His conclusion was that variation (the random production of different individuals) and selection ("survival of the fittest") are two fundamental features of life on this planet and that, together, they can account for the evolution of species.

To visualize what is so special about Darwin's idea, imagine that you are in a quandary and the situation is very complex. You have two options: 1. You can spend days analyzing the situation and trying to find the best strategy to cope with it. Or 2. You can spend only a few minutes listing ten strategies, which are more or less random and all different one from the other. In the former case, you are still thinking. In the latter case, you start applying each of the strategies at the same time. As you do so, some strategies turn out to be silly, others look promising. You pursue the ones that are promising. For example, you try ten different (random) variations on each of the promising ones. Again, some will prove themselves just plain silly, but others will look even more promising. And so forth. By trial and error (case 2.), you will always be working with a few promising strategies and possibly with a few excellent ones. After a few days you may have found one or more strategies that cope perfectly well with the situation. In case 1., you will be without a strategy for as long as you are thinking. When you finally find the best strategy (assuming that you have enough experience and intelligence to find it at all), it may be too late.

In many situations, "design by trial and error" (case 2.) tends to be more efficient than "design by a designer" (case 1.).

So Darwin opted for "design without a designer": nature builds species which are better and better adapted and the strategy it employs is one of trial and error.

The idea of evolution established a new scientific paradigm that has probably been more influential than even Newton's Mechanics or Einstein's Relativity.

Basically, evolution takes advantage of the uncertainty left in the transmission of genes from one generation to another: the offspring is never an exact copy of the parents, there is room for variation. The environment (e.g., natural selection) indirectly "selects" which variations (and therefore which individuals) survive. And the cycle resumes. After enough generations have elapsed, the traits may have varied to the point that a new species has been created. Nobody programs the changes in the genetic information. Changes occur all the time. There may be algorithms to determine how change is fostered. But there is no algorithm to determine which variation has to survive: the environment will make the selection.

Living organisms are so complex that it seems highly improbable that natural selection alone could produce them. But Darwin's theory of variation and natural selection, spread over millions of years, yields a sequence of infinitesimally graded steps of evolution that eventually produce complexity. Each step embodies information about the environment and how to survive in it. The genetic information of an organism is a massive database of wisdom accrued over the millennia. It contains a detailed description of the ancient world and a list of instructions for surviving in it.

The gorgeous and majestic logical systems of physical sciences are replaced by a completely different, and rather primitive, system of randomness, of chance, of trial and error.

Of course, one could object that natural selection has (short-term) tactics, but no (long-term) strategy: that is why natural selection has never produced a clock or even a wheel. Tactics, on the other hand, can achieve eyes and brains. Humans can build clocks, but not eyes. Nature can build eyes, but not clocks. Whatever humans build, it has to be built within a lifetime through a carefully planned design. Nature builds its artifacts through millions of years of short-term tactics. "Design" refers to two different phenomena when applied to nature or humans. The difference is that human design has a designer.

Darwinism solved the problem of "design without a designer": variation and selection alone can shape the animal world as it is, although variation is undirected and there is no selector for selection. Darwin's greatest intuition was that design (very complex design) can emerge spontaneously via an algorithmic process.

To be fair, Darwin already realized that natural selection alone was not enough to explain the evolution of very complex traits (such as the human brain itself). Thus he later introduced a second kind of selection, that, while not as popular as "natural" selection, could actually account for rapid development of complex organs in primates: sexual selection. Sexual selection is due to the different way males and females of a species behave towards reproduction: males compete for females, females choose males. Thus males and females are under pressure to develop features that not only will improve their chances of surviving in a hostile environment but will also improve their chances of reproducing with a member of the other sex. Primates are under the pressure of both natural selection "and" sexual selection (competition for survival "and" competition for reproduction). Sex is not only a footnote.

The Logic Of Replication

In 1865 the Austrian botanist Gregor Mendel, while studying pea plants, proposed a mechanism for inheritance that was to be rediscovered in 1901. Contrary to the common-sense belief of the time, he realized that traits are inherited as units, not as "blends". Mendel came to believe that each trait is represented by a "unit" of transmission, by a "gene" (a term coined in 1909 by Wilhelm Johanssen). Furthermore, traits are passed on to the offspring in a completely random manner: any offspring can have any combination of the traits of the parents.

Mendel proved that "blending inheritance" is false, that we do not "blend" the inheritances we receive from our parents. There is a unit of inheritance, the gene, and we either inherit a gene or we don't inherit it. Our eyes are either blue or brown, but not a blend of blue and brown. We are either male or female, but not a blend of male and female. (The color of the skin may be intermediate between the colors of the parents, but that is because the color is due to the sum of numerous genetic effects).

The British biologist William Bateson coined the term "genetics" in 1906. The Danish botanist Wilhelm Johannsen coined the term "gene" in 1909. In the 1920s the US biologist Thomas Hunt Morgan discovered that genes are arranged linearly along "chromosomes".

The model of genes provided for a practical basis to express some of Darwin's ideas. For example, Darwinian variation within a phenotype can be explained in terms of genetic "mutations" within a genotype: when copying genes, nature is prone to making typographical errors that yield variation in a population. In fact, in 1900 the Dutch botanist Hugo de Vries had even claimed that mutations (not natural selection) were the real cause of speciation.

In the 1920s population genetics (as formulated by the US biologist Sewall Wright and the British biologist Ronald Fisher) turned Darwinism into a stochastic theory (i.e., it introduced probabilities). Evolution became a shift in gene frequencies within a population over time. Fisher, in particular, proved that natural selection requires Mendelian inheritance in order to work the way it works. Fisher unified Darwin and Mendel (initially Mendel had even been viewed as anti-Darwin): what changes in evolution is the relative frequency of discrete hereditary units, each of which may or may not appear (more or less randomly) in successive generations. In the 1940s the two theories were merged for good in what Julian Huxley called "Modern Synthesis". In practice, the synthetic theory of evolution merged a theory of inheritance (Mendel's genetics) and a theory of species (Darwin's evolutionary biology).

Since those days, the idea of natural selection has undergone three stages of development, parallel to developments in the physical sciences: the deterministic dynamics of Isaac Newton, the stochastic dynamics of James Maxwell and Ludwig Boltzmann, and finally the dynamics of self-organizing systems. Originally, Darwin's theory was related to Newton's Physics in that it assumed an external force (natural selection) causing change in living organisms (just like Newton posited an external force, gravity, causing change in the motion of astronomical objects). However, with the formulation of population genetics by Ronald Fisher and others, Darwinism became stochastic (the thermodynamic model of genetic natural selection, in which fitness is maximized like entropy), just what Physics had become with Boltzmann's theory of gases. In the 1990s self-organizing systems provided a new model to think about the organization of life at different levels, from cells to societies.

Darwin had not explained what he set out to explain: the origin of species. The Russian geneticist Theodosius Dobzhansky came closer: what makes a species a species is sex. Different populations of the same species actually have different genomes, but only some genes need to be compatible for sexual reproduction to occur. If those genes are affected by mutations, then two populations may become incompatible and a new species is born.

In 1944 the Canadian physician Oswald Avery identified the vehicle of inheritance, the substance that genes are made of, the bearer of genetic information: the deoxyribonucleic acid (DNA for short).

In 1953 the British biologist Francis Crick and the US biologist James Watson figured out the double-helix structure of the DNA molecule. It appeared that genetic information is encoded in a rather mathematical form, which was christened "genetic code" because that's what it is: a code. The "genome" is the repertory of genes of an organism.

In 1957 Crick, by using only logical reasoning, reached the conclusion that information must flow only from the nucleid acids to proteins, never the other way around.

In 1961 the South African biologist Sydney Brenner and the French biologist Francois Jacob discovered that cells of ribonucleic acid (messenger RNA), carry the genetic instructions from the DNA to the ribosomes, the sites within a cell that manufacture proteins.

Also in 1961 Jacob and the French biologist Jacques Monod discovered the mechanism of gene regulation: genes turn each other on and off, i.e. genes are organized in a network. Jacob and Monod also noticed that the interaction among genes might explain cell differentiation (the fact that cells containing the same genetic information end up doing completely different things).

By 1966 the US chemist Marshall Nirenberg had cracked the "genetic code", the code used by DNA to generate proteins. He and the Indian biologist Har Gobind Khorana discovered how the four-letter language of DNA is translated into the twenty-letter language of proteins (the DNA is made of four kinds of nucleotides, proteins are made of twenty types of aminoacids).

In the 1980s we started deciphering the genome of different animals, including our own.

The Logic Of Evolution

Evolution is about a pattern (in particular, a string of DNA, but it could also be some other pattern) and it involves the following steps:
- Reproduction. Copies are made of the pattern.
- Variation. Random errors appear in the copies and yield variants.
- Selection. The environment selects which variants survive.

These simple steps cause continuous mutations of the pattern. Each generation copes better with the environment.

In addition, other factors may accelerate evolution: sex accelerates evolution; learning accelerates evolution.

Universal Selectionism

The thesis of the US psychologist Gary Cziko is that there is a universal process of Darwinian evolution that is responsible for knowledge at all levels, and not only at the biological level.

Cziko relates "knowledge" to the fitness of living beings, to their adaptation to the environment. Knowledge as the product of the interaction with the environment (and as the necessary cause for survival in that environment) implies that all knowledge is created through a Darwinian process of blind variation coupled with environmental selection.

Cziko observes that there are really two kinds of fitness: living beings are adapted to their environment when they are born, and living beings are capable of adapting to changes in their environment during their lifetime. A theory of fitness has to deal with both forms of fitness, the one that has been shaped over the centuries and become part of a species' identity and the other that is shaped over an individual's lifetime and becomes part of the individual's identity ("instinct" and "learning").

The Austrian ethologist Konrad Lorenz saw instinct as having been shaped by blind variation and natural selection: it is "knowledge" acquired over millennia that (in modern terms) is now encoded in the genome of a species. However, the same behavior does not yield the same outcome unless the environment remains exactly the same; which, in general, it doesn't. William James noted that a living being is capable of achieving consistent goals using (slightly) different behaviors. Individuals can "adapt" to circumstances.

Examples of "ontogenetic" adaptation are the muscles that get bigger the more they are used and the immune system, that "learns" what antibodies to make based on which ones are "used" to fight antigens. The immune system is particularly effective in its job. However, it operates on an absolutely blind basis: it creates all the time a lot of different kinds of antibodies hoping that, when attacked, at least one will work against the invader. It is the diversity of its army of soldiers (and the fact that they are permanently available) that makes it effective in fighting the enemy, not a careful training of each soldier and a timely deployment of them. It's diversity and continuity that matter. And they are due to a process of blind variation, not to a process of careful engineering. (Another key feature for the proper functioning of the immune system is, of course, that it produces only antibodies that destroy antigens and no antibodies that destroy body cells; in other words, it is capable of distinguishing self from nonself).

The "blindness" of selectionist processes turns out to be an advantage in other ways as well. There are several examples of "functional shifts", i.e. of parts that evolved for a purpose but then ended up being used for a different purpose, simply because it worked (what the US biologist Stephen Jay Gould termed "exaptation"). The current use of an organ or behavior does not necessarily explain its origin. It may well be that it originated for a different function.

The other major example of "ontogenetic" adaptation is the brain that is shaped not only by the genes but also by experience. Interaction with the environment "selects" which synapses are useful and eliminates the ones that are useless.

Drawing from Konrad Lorenz's evolutionary epistemology of 1941, Cziko views knowledge as adaptation of the brain to the environment. A-priori knowledge (the innate knowledge of entities such as time and space) is the product of the biological evolution of the human brain. During our lifetime the senses provides us with true information about the environment because they have been selected over the millennia based on their usefulness. Therefore the world is not an illusion, and we know it by adapting to what it is.

Cziko observes that "fitness" has to do with "purpose": there is fit when the structure of an organism serves a purposeful function. Animal behavior is purposeful and changes the environment that operates on the animal's behavior. Stimuli influence responses, but responses also influence stimuli. This is William Powers' "perceptual control theory", according to which behavior controls perception as much as perception determines behavior. A control system is as blind as the immune system that creates an army of antibodies. Nonetheless, a control system exhibits a behavior that appears to be "purposeful". It is, in turn, "controlled" by higher-level control systems. An organism is ultimately a hierarchy of control systems, each of which senses something in the environment and tries to control it. Instinctive behavior is the result of the interaction between control systems that have internal goals. Each control system must have survival value if it is still part of an organism. In a sense, there is no learning: there is just the blind functioning of a network of control systems. In another sense, that "is" precisely what we call "learning": a control system at work. When something changes in the environment, the control system senses it and needs to restore its internal goal. It does so by triggering random responses and rewarding the ones that move it closer to its goal. A hierarchy of control systems can create the illusion of learning and of intelligence (as in Valentino Breitenberg's progressively complex robots).

Cziko argues that there is a Darwinian selection not of behaviors but of control systems. Knowledge is the result of a hierarchy of selectionist processes, starting with the biological one studied by Darwin.

"Learning" is necessary because animals need to adapt to changing environments. The brain and the immune system allow animals to find food and to fight lethal viruses. Humans have also developed a higher form of "learning" that consists in cultural knowledge. Cziko shows that it too obeys a process of blind variation and selective retention. Even when we learn something from somebody else we are simply interacting with the environment (the "somebody else") and fine-tuning our knowledge based on that interaction. Both technological and scientific evolution, for example, are due to a number of different "trial and error" processes.

Cziko applies "universal selectionism" to a number of different fields. The most interesting is language. Following the US linguists Elizabeth Bates and Brian MacWhinney ("Competition, variation, and language learning", 1987), Cziko shows that a selectionist approach can well complement Chomsky's nativism to explain how children learn language. There might be innate linguistic skills in the brain (selected over the millennia by evolution) but children learn a language by the same "trial and error" process that Nature employs everywhere. Children try words and sentences and reinforce the ones that "work", just like the brain and the immune system try synapses and antibodies and reinforce the ones that work (or, better, the ones that work are reinforced by the positive outcome).

Genes

An organism is a set of cells. Every cell of an individual (or, better, the nucleus of each cell) contains the DNA molecule for that individual, or its "genome".

"Polymerizing" is the process by which molecules form chains, therefore called "polymers". The polymer of life is formed by molecules of four kinds (four "nucleotides").

A DNA molecule is made of two strings, or "strands", each one the mirror image of the other (in the shape of a "double helix"). Each string is a sequence of "nucleotides" or "bases", which come in four kinds (adenine, guanine, cytosine, thymine). These four bases are paired together (adenine is paired with thymine and cytosine is paired with guanine). Each nucleotide in a string is "mirrored" in a nucleotide of the other string. Each strand of the helix acts therefore as a template to create the other template. Nucleotides are the elementary unit of the "genetic code". In other words, the genetic code is written in an alphabet of these four chemical units.

Cells split all the time, and each new cell gets one of the two strings of DNA of the original cell, but each string will quickly rebuild its mirror image out of protoplasm. This process is known as "mitosis". Each cell in an individual has almost exactly the same DNA, which means that it carries the same genome.

The genome is made of genes. A gene is a section of the DNA molecule which instructs the cell to manufacture proteins (indirectly, a gene determines a specific trait of the individual). Genes vary in size, from 500 bases long to more than two million bases (long genes tend to have just a very long waste).

The most abused metaphor in biology is that genes represent a program that results in some behavior (the "digital gene" metaphor). In reality, the behavior of genes is not so linear as the digital metaphor imply. Genes tend to work in communities of genes: it is not always clear what a gene does. Some genes are used for more than one chore (the "housekeeping genes"). And some genes do not encode discrete values, but continuous values. Many genes, in other words, are not digital at all. And the genome is not a sequential program, that is executed mechanically one gene after the other. It is more like a network of genes that "regulate" each other. The genetic "program" behaves more like a network of switches.

The DNA is organized into chromosomes (23 pairs in the case of the human race) which are in turn organized into genes. The human genome has 3 billion base pairs of DNA.

This means that each cell contains three billion bases of DNA, which is a string of genes about 2 meters long. If we multiply for all the cells in the human body, we get a total length of genetic material which is about 16,000 times the distance between the Earth and the Moon.

The way offspring is designed is simple: male sperm and female eggs carry only 23 chromosomes (instead of the 46 that each body cell contains) and when they join they generate a 46-chromosome embryo. The embryo therefore contains some of the chromosomes of the father and some of the chromosomes of the mother. (As Mendel discovered, the embryo does not contain a "blend" of the mother and the father, but rather some of the mother's attributes and some of the father's attributes).

(Notable among the human chromosomes are the X and Y chromosomes, that are responsible for determining the sex of the offspring. Reptiles do not have genes that decode sex: sex is determined by environmental conditions, mostly the incubation temperature, not by genetic information. The X and Y chromosomes were acquired by mammals much later in evolution. The Y chromosome is only one third the size of the X chromosome, and the Y chromosome has disappeared in several mammals. Human males have one X and one Y chromosome, while females have two X chromosomes).

All living organisms use DNA to store hereditary information and they use the exact same code (the "genetic" code) to write such information in DNA: the genome of an individual is written in the genetic code. It is inappropriate (although commonplace) to refer to the "genetic code" of an individual, as all living things on this planet share the same genetic code. The genetic code is a code, just like the Morse code. It specifies how nucleotides (through a "transcription" of the four nucleotides into ribonucleic acid, or RNA, and a translation of RNA into the twenty aminoacids) are mapped into aminoacids, which in turn make up proteins, which in turn make up bodies. Different genomes yield different bodies. But they always employ the same genetic code to carry out this transformation.

The genetic code is the code used by Nature to express a set of instructions for the growth and behavior of the organism. Each individual is the product of a genome, a specific repertory of genes written in the genetic code. The genome defines the "genotype" of an organism. Genotype is the "genetic makeup" of the organism. The organism itself is the "phenotype". Phenotype refers to how the genetic makeup is expressed in the body (the physical expression of a gene). The genotype is the repertory of genes of an organism; the phenotype is the physical manifestation of the genotype (the "body").

"Sequencing" the genome refers to the process of identifying the genes.

Humans have about 20,000 genes (out of 3.2 billion DNA units). That is a relatively low number for the complexity of the human body (only six times more than the Escherichia Coli bacterium).

A single gene can often be responsible for important traits. For example, chimpanzees share 98.6% of the human genome, but there is hardly a single trait in common between the two species. 98% of the human genome contains the same DNA found in most other vertebrates. The roundworm has 19,000 genes, just a third less than humans. But a single gene can make a huge difference and very similar genetic programs can differ wildly in phenotypic (bodily) effects. In other words, the relationship between genome and phenotype is nonlinear: the genotypes of humans and chimps differ by only 1.4%, but the difference in the corresponding phenotypes is much more striking than a mere 1.4% (at least from the admittedly biased viewpoint of us humans). To be fair, the 20,000 genes represent only 1.5% of the genome, the rest being "junk", i.e. chemicals that don't seem to encode any instruction.

Some of those genes that humans share with chimps, incidentally, have been around for millions of years, and humans share them with bacteria. As the British biologist Steven Jones wrote, "everyone is a living fossil".

The smallest genome that is known is the genome of the Mycoplasma Genitalium: 470 genes. One could wonder what is the smallest amount of genes that is required to have life.

The Matter of Life

Life on Earth is based on the element carbon. Carbon can bond with oxygen, hydrogen and nitrogen because of its four valence electrons. Both proteins and nucleic acids (DNA, RNA) are made of carbon, and energy is stored in the form of carbohydrates.

Energy transformations are linked with the dual processes of oxidation and reduction. Oxidation and reduction reactions occur in pairs: in order for a substance to be reduced another substance must be oxidized. When an atom is oxidized, it loses electrons; and when it is reduced, it gains electrons. Oxidation is a process by which a carbon atom gains bonds to (usually) oxygen. Reduction is a process by which a carbon atom gains bonds to (usually) hydrogen. Molecules gain oxygen or lose hydrogen in an oxidation reaction; lose oxygen or gain hydrogen in a reduction reaction.

Fritz Lipmann ("The metabolic generation and utilization of phosphate bond energy", 1941) discovered that Adenosine Triphosphate (ATP), a high-energy phosphate and therefore a source of readily available energy, and showed that it constitutes the main fuel for many biochemical processes and, in particular, is the main energy carrier in cells. He had basically found the connection between the metabolism and the physics of biological systems, i.e. the chemical process that makes life possible.

Organic molecules are based on carbon, and energy transactions are based on phosphorus.

Oxidation-reduction energy is pervasive in nature. and (relatively rare) Phosphate-bond energy is relatively rare in nature. Life owes its existence to the fact that at some point the two met.

Genetic Fossils
Genomes have confirmed the theory of evolution. In the 1950s the Italian biologist Luigi Cavalli-Sforza first had the idea that one could use genetic information to trace the genealogical tree of species. Genomes share common parts and different species are determined by the branching out of the other parts. The genealogical tree of living beings is carefully reflected in the structure of their genomes. The genome of a species is almost a "memory" of that species' evolutionary journey. Most human genes, for example, date back to primitive organisms, and they are shared by all modern animals that descend from those organisms. Only a few can be said to be truly "human".

Basically, an organism's DNA is a record of its evolutionary past. Each living organism "is" a fossil. The same principle helped biologists such as Allan Wilson in the 1980s study the evolution of humans. He focused on mitochondria, the part of the cell that converts sugar into energy. They have their own DNA. This DNA can be used as a "molecular clock" by estimating the number of its mutations.

Because mitochondria are inherited only from the mother, they can only be used to construct a matrilineal genealogical tree. Thus it was derived that all living humans are descendants of one woman who lived about 150,000 years ago. The molecular clock for the patrilineal genealogical tree is a piece of the Y chromosome, which is inherited only by sons from their father. Thus it was derived that all living humans are descendants of a man who lived about 60,000 years ago.

Differentiation

A body is made of cells. Every single cell in the same body contains roughly the same genetic information (barring copying mistakes). However, each cell ends up specializing in a task, depending on where it is located: a heart cell will specialize in heart issues and not, say, liver issues, even though the genetic information describes both sets of issues. A muscle cell is a muscle cell, even though it is identical to a liver cell. This is the phenomenon of "cell differentiation", by which each cell "expresses" only some of the genes in the genome, i.e. only some of the possible proteins are manufactured ("synthesized").

The human body has about 265 different cell types.

Differentiation seems to be regulated by topology: depending on where a cell is, it exchanges energy (which is information) with some cells rather than others. Neighboring cells "self-organize". How cells develop to be what they will be within the body is probably determined by a regulatory mechanism: instead of each cell being "told" by the genes what to become, cells interact among themselves; and the body is the emergent outcome of their interaction. This is an efficient way to produce complex bodies: the genome does not need to specify where each of the 100 trillion cells must go. The difference between humans and chimps is caused by a mere 1.6% of the genome: it is the interactions among cells that greatly amplify that 1.6%. Another advantage is that the body can repair itself: if a cell is damaged, interaction among cells yields a new configuration that makes that damage irrelevant. The price that the organism pays is a long period of "incubation" during which the body "develops" (and it is vulnerable).

A puzzling feature of genomes is that they contain far more useless junk than useful genes. The human genome, in particular, contains about 95% junk, in between genes.

Epigenesis

The process of "epigenesis" is the process by which the genotype is turned into the phenotype.

DNA is translated into another kind of polymer, RNA (ribonucleic acid), which is also a four-letter code (adenine, guanine, cytosine, uracil) while RNA, in its turn, is translated into yet another kind of polymer, the protein, whose units are aminoacids. There are twenty kind of aminoacids, so technically a four-letter code is translated into a twenty-letter code.

The way this works is by triplets of DNA units: three DNA units (which can each be of four different kinds, for a total of 4x4x4=64 combinations) is translated into one of the twenty aminoacids (some triplets generate the same aminoacid).

But the translation is more complex as DNA is a one-dimensional structure (a string) whereas a protein is a three-dimensional structure (it's the stuff that our flesh and bone and blood are made of). So the one-dimensional string of instructions of the DNA is used to determine the three-dimensional shape of a protein.

In summary, the DNA is the sequence of instructions for building molecules called proteins, and proteins are manufactured of amino acids, whose order is determined by the DNA. Note that our genome has only 20,000 genes, but our body has 100 trillion cells.

As far as the individual goes, we know that her genome is a synthesis of the genome of the parents plus some random shuffling. But it is not clear yet how much of the final individual is due to the genome and how much to the interaction with the environment. For example, the genome may specify that a muscle must grow between the arm and the trunk, but exercise can make that muscle bigger or smaller. For example, the genome may determine some psychological characteristics of the individual, but study, meditation and peer pressure can alter some of them. The British biologist William Bateson thought that only the genome mattered: we are machines programmed from birth. The US psychologist John Watson, on the other hand, thought that conditioning could alter at will the personality of an individual: it all depends on experience, the instruction contained in the genome is negligible.

There is also a subtle difference between which genes are in the genome and which genes are actually "expressed". For example, the Israeli physician Moshe Szyf ("Maternal Care Effects On The Hippocampal Transcriptome And Anxiety-Mediated Behaviors", 2005) found evidence that the early experience of the child affects the future psychological life of the child not only because it is stored in memory but also because it determines how some genes will be expressed. Szyf observed physical differences in the hippocampus of rats that account for differences in behavior, and he argued that those differences were caused by the way their mothers raised them. Rats who were raised in similar ways by their mothers tend to have the same kind of hippocampus. He credited this development to the expression of some genes as opposed to others. Maternal care seems to affect the chemistry within the cell that determines if and when those genes are expressed.

The role of RNA is probably underrated. Protein-encoding genes might be in the minority. Many different kinds of RNA exist and some kinds of RNA regulate the life of many protein-encoding genes. The number of protein-encoding genes seems to be mostly the same for all animals, from flies to humans (in the range of 20-30,000). However, the number of genes whose RNA performs other functions vary wildly among species. RNA acts as a simple "messenger" only in simpler organisms. RNA acts more like a "manager" in complex organisms, i.e. its "regulating" activities are much more widespread.

The Cell
The cell itself is the elementary unit of the body (in the case of bacteria, one cell is the whole body), but it has its own structure. Its behavior is driven by the instructions of the DNA, but its function are roughly to transform nutrients into energy (generically, metabolism), to carry out some function thanks to that energy, and to reproduce (mitosis).

Each cell is limited by a membrane and is supported by a skeleton called a "cytoskeleton". The cytoskeleton is made of a protein called "tubulin", which forms filaments called "microtubules". Most living beings are eukaryotes: their cells contain a nucleus that contains the genetic material (DNA and RNA). The nucleus is the information storage of the cell. It is also the place where DNA is transformed into mRNA. The cell contains many ribosomes, each of which is a machine that manufactures proteins based on the instructions received from the nucleus in the form of mRNA. There is also genetic material in some organelles of the cell called "mitochondria". Mitochondria are the cell's "power plant," because they convert nutrients into energy, creating the nucleotide that is the "molecular currency" of intracellular energy. Mitochondria have their own DNA, separate from the cell's main DNA. Mitochondria are also self-replicating.

Prokaryotes (such as bacteria) are made of simpler cells that do not have a nucleus.

Mutation as Destiny
In reality, the process of copying DNA is not so smooth. When a cell splits, its DNA is copied to the new cells but the copying process (for whatever whim of nature) is prone to "error" (or, at least, to loss of information). In other words, genes mutate all the time inside our bodies. These mutations may cause fatal diseases (such as cancer) and they are responsible for death.

Mutation is what causes aging and death. Millions of cells divide each second and a copy of DNA is likely to carry some mistake, which means that the older we are the more chances that serious mistakes have been made and that our genetic instructions are no longer rational.

Mutation is also the whole point of sex, and this turns out to be the mirror story of death. Sex is the antidote to the genetic deterioration due to the imperfect copying process. The human race would rapidly degenerate without sex: each individual would pass on genes that have already lost part of their information through so many million internal copies. Sex is what makes the paradox possible, and almost inevitable: individuals decay, but the race progresses. Because sex recombines the genes of the parents, it can produce both better and worse (genetically speaking) individuals, and natural selection will reward the better ones. The long-term outcome of sex is that it is more likely that better future individuals are produced from the deterioration of present individuals.

Last but not least, mutation is what drives evolution (evolution is variation and natural selection).

Mutation sounds like the god of genetics.

The problem is that mutation is random. Evolution occurs by accident, by "genetic drift": by chance and time.

Mutation is not everything, though. Mutation requires natural selection in order to yield evolution.

Inheritance involves genes and environment working together. Diseases which are dormant in our genes, for example, may be sparked off by environmental conditions. Diet is as important as genes in the development or the prevention of a disease. And pollution is as important as genes to the development of cancer. And so forth.

Chance and the environment determine how we evolve. The only party that does not have a saying in this process is... us.

The Origin of Evolution

Fundamental to Gregor Mendel's theory is the distinction between the appearance of an organism (its "phenotype"), which turns out to be a blend of the appearances of its parents, and the physical state of the factors inherited from each parent (the "genotype"), which remain unmixed. The physiology of development fuses, at the level of the whole organism, the information of heredity, which is still kept separated at the genetic level. The two fundamental laws of heredity are that, first, the factors that are passed from parent to offspring (which today we call "genes") maintain their individuality despite their interaction with other genes in the development of the organism, and that, second, gene segregation allows for the reappearance of a variation in later generations of offspring. From these considerations Mendel had the intuition that heredity is based on a discrete (rather than continuous) entity, just like Physics is based on elementary particles. That entity was the gene. What is truly inherited is not the "traits": it is the genes.

(Darwin, incidentally, believed that traits were transmitted from parent to offspring through blood).

Mendel also found that new variation will not be diluted by the process of mating but will always be available for selection, a fact that explains why a population variation is not immediately destroyed by selection itself. The antithetical properties of heredity and variation are dual aspects of the same process: the actual variation among members of the same generation explains the transmission of similarity across generations.

In our century, population genetics showed that Darwin's theory (that change occurred by the natural selection of many minute variations) and Mendel's theory (that change occurred suddenly, by mutation) were complementary: changes occur in the frequencies of genes.

Modern evolutionary genetics stems from the merging of those two traditions, the Darwinian and the Mendelian, both of which take variation as the crucial aspect of life. The Darwinian view can be summarized as "evolution is the conversion of variation between individuals into variation between species".

The paradox is that Mendelian theory dictates the frequencies of genotypes as the appropriate genetic description of a population, whereas variation is much more important. As the US biologist Richard Lewontin put it, "what we can measure is uninteresting and what we are interested in is unmeasurable".

The Steps Of Life
Life evolved through momentous leaps forward.

As the Israeli physicist David Deutsch described it, once replicators were born, they joined forces in self-replicating groups. Such groups are organized in a way that each member contributes to the chemical reactions that allow the whole to replicate itself (with all its members). That was the birth of the first living organism. "Genomes are group of genes that are dependent on each other for replication". The genetic code itself (the way to encode all of this) had to evolve until it reached a point beyond which it did not need to evolve anymore (it hasn't evolved for billions of years) while still allowing for organisms to evolve. The code, that had originally encoded just a single-celled organism, stopped evolving, but it was now powerful enough ("universal") that it could encode a virtually infinite range of (multi-cellular) organisms. That is the power of representation systems when they are universal. They can describe a lot more than what they were originally built for.

First, reproduction occurred: an organism became capable of generating another organism of the same type. Then sexual reproduction occurred, in which it took two organisms to generate an organism of the same type. Then multi-cell organisms appeared, and organisms became complex assemblies of cells. Fourth, some of those cells developed into specialized organs, so that the organism became an entity structured in a multitude of more or less independent parts. Fifth, a central nervous system developed to direct the organs. And, finally, mind and consciousness appeared, probably originating from the same locus that controls the nervous system.

The Origin of Life

Hypotheses abound on how life originated. Most theories analyze the ingredients of life and speculate how they may have been generated by the Earth's early activity.

It was in 1952 that a young US physicist, Stanley Miller, working with Harold Urey, advanced the idea that the first molecules of life (including aminoacids, the building blocks of proteins) were formed accidentally by the Earth's early volcanism and then triggered into reproducing systems by the energy of the sun and lightning strikes. His calculations of how lightning may have affected the Earth's primitive atmosphere gave rise to the quest for the experiment that would reproduce the birth of life in a laboratory (with hints of Frankenstein and all the rest). One catch remained, though: the product of Miller's prebiotic chemistry would still be inactive chemicals.

Miller simply revised a theory of chemical formation of life that dates back to the Russian chemist Alexander Oparin, who in 1924 first proposed that life could have been induced in the primeval soup.

Autocatalysis

Since the pioneering work conducted in the 1960s by the German physicist Manfred Eigen ("Self organization of matter and the evolution of biological macro molecules", 1971), autocatalysis has been a prime candidate to explain how life could originate from random chemical reactions. Autocatalysis occurs when a substance A catalyzes the formation of a substance B that catalyzes the formation of a substance C that... eventually catalyzes the formation of A again. At the end of the loop there is still enough A to restart it. All the substances in this loop tend to grow, i.e. the loop as a whole tends to grow. Life could have originated precisely from such a loop, in which case the chances that the right combination of chemical reactions occurred at the right time is much higher.

An autocatalytic set is a group of proteins that reproduces itself.

The power of this hypothesis is that "autocatalytic cycles" exhibit properties usually associated with life: metabolism and reproduction. If two such cycles occur in the same "pond", they will compete for resources and natural selection will reward the "best" one.

The German patent lawyer Gunter Waechtershauser ("Before enzymes and templates: theory of surface metabolism", 1988) improved on that model by explaining how the first forms of life could have synthesized their own vital chemicals rather than absorbing them from the environment, i.e. how a metabolic cycle could have started. Unlike Stanley Miller, Waechtershauser speculates that prebiotic reactions occurred not in water but on the ground. At high temperatures, chemicals bound to a metallic surface are much more likely to mix and form the complex molecules which are needed for life. Particularly, iron sulfide (a very common mineral on the Earth) could have been a catalyst of chemical reactions that created the biochemistry of living cells. He proved that peptides (short protein chains) could be created out of a few given aminoacids. The next step in the chain would be the emergence of RNA, that he considers a predecessor to DNA. Waechtershauser's emphasis is on "autocatalysis" (in general, as a process that is fast enough for yielding dramatic consequences) and on the ability of minerals in particular to catalyze the right reactions. Life would be but the natural evolution of a primitive chemical cycle that originally arose on an iron-sulfur surface.

The US chemist Melvin Calvin was perhaps the first to suggest that "autocatalytic" processes can make life more likely by speeding up the manufacturing of the basic ingredients.

The US biologist Stuart Kauffman also advanced a theory of how life may have originated from autocatalysis. He refutes the theory that life started simple and became complex in favor of a scenario in which life started complex and whole due to a property of some complex chemical systems, the self-sustaining process of autocatalytic metabolism. When a system of simple chemicals reaches a certain level of complexity, it undergoes a phase transition: the molecules spontaneously combine in an autocatalytic chemical process to yield larger molecules of increasing complexity and catalytic capability. In other words, as the system gets more complex, the chances that it contains a component and its catalyzer increase rapidly. Even if the proteins are chosen randomly, when there are enough of them, there is a chance that some of them form an autocatalytic set. Self-replication arises out of a simple statistical fact. Life, according to Kauffman, is but a phase transition that occurs when the system becomes complex enough.

According to Kauffman, life is vastly more probable than traditionally assumed. And life began complex, not simple, with a metabolic web that was capable of capturing energy sources.

Self-organizing principles are inherent in our universe, and Kauffman views life as a direct consequence of self-organization. Therefore, both the origin of life and its subsequent evolution were inevitable.

The US scientist Michael Conrad ("The Fluctuon Model of Force, Life, and Computation", 1993) developed a unified model of Quantum Physics and General Relativity, the "fluctuon model", according to which Physics is inherently biased towards self-organizing processes. He argued that life-like features stem from Quantum Physics and General Relativity themselves, and that life is therefore a relatively trivial consequence of the evolution of the universe.

Panspermia

Comets are providing another option: that life may have come from other parts of the universe. It was the Greek philosopher Anaxagoras (fifth century BC) who first speculated that life may have been dispersed as seeds in the universe and eventually landed on Earth ("panspermia") and it was the Swedish chemist Svante Arrhenius ("The Propagation of Life in Space", 1903) who gave this theory scientific credibility.

Spanish biochemist John Oro ("Comets and the formation of biochemical compounds on the primitive Earth", 1961) hypothesized that all building blocks of life were brought to Earth by comets, a theory later popularized by the Belgian astrophysicist Armand Delsemme. Organic material (from water to methyl alcohol, and even forerunners of DNA's aminoacids) has been found in the galactic clouds that float among the stars of our galaxy. Interstellar matter seems to be rich in molecules that are needed to create life. Trillions of comets wander through the solar system, and they occasionally approach the Earth. They are soaked with the organic dust picked up from the interstellar void. In other words, comets may have their own role in the vast drama of life, sowing the seeds of life on all the planets they intersect. Comets have been found to contain many if not all the ingredients necessary for life to originate. (Incidentally, comets have been found to carry ice, and no theory of the development of the Earth can account yet for the enormous quantity of water contained in the oceans, unless the water came from somewhere else).

Also, left-handed aminoacids (the kind that life uses) were found in the meteorite fragments that showered Australia in 1969 (including some aminoacids unknown on Earth).

If aminoacids are of extraterrestrial origin and Wachtershauser's mineral-based chemistry can produce biological compounds, the chain that leads from dead matter to living matter would be completed. But life is also capable of reproduction and inheritance. Moreover, Wachtershauser's model requires high temperatures, whereas four of the five main components of DNA and RNA (adenine, uracil, guanine, cytosine) are unstable at those temperatures.

Thermosynthesis

The Dutch chemist Anthonie Muller showed that "thermosynthesis" is a viable alternative to explain the origin of life ("Thermoelectric energy conversion could be an energy source of living organisms", 1983). Muller points out that life probably originated in conditions where photosynthesis and chemosynthesis (getting energy from light and food) were unfeasible, simply because there were not enough life and food. If life originated in an underwater volcano covered with ice, neither light nor food were abundant. What was abundant was a temperature difference. This "gradient" of temperature would cause convection currents, that would drag the early forms of life up and down in thermal cycles, from hot to cold and back to hot. The larger the temperature difference, the stronger the convection currents, the faster the thermal cycles, the more efficient the energy production. Heat was therefore the main source of energy, and heat was coming from the environment.

Photosynthesis and chemosynthesis do yield much more power, but thermosynthesis was simply the only feasible form of energy production. The early living cells were basically built around "heat engines". Some of their enzymes or membranes worked essentially as heat engines.

In a steam engine, for example, water is thermally cycled: water is heated until it turns into steam; the steam expands and performs work; the steam loses its energy and returns to liquid form; and the cycle resumes.

In a thermosynthetic cell, a protein is thermally cycled in a similar manner: it is heated until it turns into a more fluid state; this generates work in the form of ATP (the chemical which is the energy source for almost all physiological processes) while the protein returns in its original state; and the cycle resumes.

Life Before Life
Other theories focus on the replication mechanism, which doesn't necessarily require organic matter to occur.

For example, the British chemist Graham Cairns-Smith argued that the first living beings were not carbon composts but clay crystals, i.e. minerals. He agrees with skeptics who think that the birth of the first cell is just statistically impossible (he calculated the probability of all the events required to create a DNA molecule and concluded that there wasn't enough matter or time in the universe to achieve it). However, rather than invoking an external force, Cairns-Smith thinks that the most plausible explanation is in the other direction: life is not the towering accomplishment of Nature, but a mere leftover from something bigger that pre-existed. He compares it to some unlikely rock structures that can be found in natural parks: how could chance create such equilibrium-defying structures? They were actually part of a much bigger structure that crumbled to pieces. It was relatively easy for them to be created as part of the bigger structure. Now that the bigger structure is gone, they look surreal and unlikely. Ditto for life: Cairns-Smith believes that life is merely what is left of something that was much more likely to arise than a mouse or a bird. In his opinion, life is the remnant of a mineral process. Life's ancestors were self-replicating patterns of defects in clay crystals. One day those patterns started replicating in a different substance, carbon molecules. In a sense, Cairns-Smith wants to extend evolution to the pre-biotic world, to the world before life was born. (But these molecules are still purely self-replicating entities: it remains unexplained how they started growing bodies...) Basically, Cairns-Smith argued that evolution came first, and life came afterwards, as an accidental side-effect.

Synthetic self-replicating molecules that behave like living organisms have been crafted in the laboratory. The US chemist Julius Rebek ("Self-replicating system", 1990) recreated artificially the principles of life postulated by the biologist Richard Dawkins: "complementary" molecules (ones that fit into each other by way of spatial structure and chemical bonds) and even self-complementary molecules.

The US chemist Jeffrey Wicken showed that the thermodynamic forces underlying the principles of variation and selection begin to operate in prebiotic evolution and lead to the emergence and development of individual, ecological and socioeconomic life. He treated the prebiosphere (i.e., the Earth before life emerged) as a non-isolated closed system in which energy sources create steady thermodynamic cycles. Some of this energy is captured and dissipated through the formation of ever more complex chemical structures. Soon, autocatalytic systems capable of reproduction appear. Living systems, according to his theory, are but "informed autocatalytic systems".

British biochemist Nick Lane pointed out that the "primordial soup" didn't have enough energy to start the chemical reactions necessary to produce life's chemistry. He argued in favor of life coming from the bottom of the ocean, in hot mineral-drenched waters. Natural gradients originate when these solutions cool down, and these gradients may have been the precursors of cell membranes.

Life And Heat

Whatever the mechanism that created it, the progenitor of all terrestrial life, four billion yeas ago, was able to tolerate the extreme heat conditions of the time (a few hundred degrees or even a thousand). As a matter of fact, if we walk backwards up the phylogenetic tree (the tree of species), we find that genetically older organisms can survive at higher and higher temperatures. Thermophiles (the microbes that live at temperatures of 70-80 degrees) are living relics of the beginnings of life on Earth.

Based on such a phylogenetic tree, the US biologist Carl Woese proposed a classification of living creatures in which thermophiles (or "archaea", first discovered in 1964 by the US biologist Thomas Brock) are different both from eukaryotes (in which DNA is held by a nucleus) and prokaryotes (in which DNA floats free in the cells of bacteria): in thermophiles, DNA floats free (like in prokaryotes) but resembles the DNA of eukaryotes. Thermophiles can be found underground: some have been retrieved from 3 km beneath earth. An archaea has about two million base pairs of DNA (a human cell has about three billion).

The Australian physicist Paul Davies retraced the history of life on Earth and concluded that it began inside the Earth, with microbes that lived several kilometers under the crust of the Earth. His reasoning was that the surface of the Earth and the oceans were just too unstable and dangerous for life to appear and survive. Furthermore, the record of genes seems to prove that the ancestor of all life forms lived underneath the Earth's surface at very high temperatures.

Surprisingly, very little has been made so far of a discovery due to the French chemist Louis Pasteur in the 19th century: that living systems prefer molecules with a certain handedness (all proteins are made of L-aminoacids, i.e. aminoacids that twist light to the left, and genetic material is made of D-sugars). This molecular asymmetry is, actually, the only difference between the chemistry of living and of inanimate matter. That "handedness" (or, better, "chirality") did not exist in the inanimate matter that eventually became animate. The US chemical biologist Gerald Joyce ("A cross-chiral RNA polymerase ribozyme", 2014) has created in the laboratory an RNA-copying enzyme (a ribozyme) that works in a chemically symmetrical world (i.e., in a soup of both left- and right-handed matter), hinting at the possibility that chirality emerged after life first evolved.

The Origin of Replication

The mystery of the origin of genes is particularly challenging because a gene is such a complicated structure and is unlikely to evolve spontaneously.

The US biologist Walter Gilbert noted that most of a person's DNA does not code genes but what appears to be gibberish, and even the part that is code is distributed in fragments (or "exons") separated by useless pauses (or "introns"). In his opinion the first genetic material was made of exons, that symbiotically got together and formed new, more complex genetic material. Introns are not random leftovers, but sort of gluing elements from the original material. In a sense, his theory points to the possibility that the gene is not the ultimate unit, but exons are.

Attention has been focusing on RNA since RNA has been shown to be a self-replicating molecule that can act as its own catalyst. DNA cannot make copies of itself, and proteins cannot create themselves. They both depend on each other. But (some kind of) RNA can act as its own enzyme (i.e., its own catalyst). Therefore, RNA is capable of replicating itself without any need for proteins.

Stanley Miller proposed that the first living creatures may have been able to synthesize protein and reproduce without the help of the DNA, depending solely on RNA to catalyze their growth and reproduction. The US chemist Thomas Cech had already proven (in 1982) that RNA molecules alone can induce themselves to split up and splice themselves together in new arrangements. It is also chemically plausible that all four RNA nucleotide bases could have been created in nature by ordinary atmospheric, oceanic and geological processes. Miller's theory, though, requires that life be born in lukewarm water, not the very high temperatures of thermophiles.

The German physicist Manfred Eigen induced RNA molecules to replicate by themselves, thereby lending credibility to the hypothesis that RNA came before DNA and that the first forms of life employed only RNA. Eigen's experiments with "autocatalytic cycles" involving RNA showed that, under suitable conditions, a solution of nucleotides gives rise spontaneously to a molecule that replicates, mutates and competes with its progeny for survival. The replication of RNA could then be the fundamental event around which the rest of biology developed. Eigen speculates that the genetic code was created when lengths of RNA interacted with proteins in the "primordial soup". First genes were created, then proteins, then cells. Cells simply provide physical cohesion. Cells first learned to self-replicate and then to surround themselves with protective membranes.

The US physicist Freeman Dyson emphasized that one cannot consider life only as metabolism or only as replication. Both aspects must be present. Therefore, we must look not for the origin of life, but for the origin of replication and for the origin of metabolism. Since it is unlikely that both metabolism and replication occurred at the same time in one of the primitive organic molecules, Dyson thinks that life must have had a double origin. It is more reasonable to assume that life "began" twice, with organisms capable of reproduction but not of metabolism and with (separate) organisms capable of metabolism but not of reproduction, and only later there arose a mixture of the two by some kind of symbiosis: organisms capable of both reproduction and metabolism.

Dyson's idea is that organisms that could reproduce but not replicate came first. The most elementary form of reproduction is simply a cell division: two cells are created by dividing a cell into two. Replication implies that molecules are copied. Reproduction with replication implies that the new cells "inherit" the molecules of the mother cell. Replication became a parasite over metabolism, meaning that organisms capable of replication needed to use organisms capable of metabolism in order to replicate. First proteins were born and somehow began to metabolize. Then nucleic acids were born and somehow began to replicate using proteins as hosts.

The two organisms became one thanks to a form of symbiosis between host and parasite. Dyson borrows ideas taken from Manfred Eigen (who claims that RNA can appear spontaneously) and Lynn Margulis (who claims that cellular evolution was due to parasites). Basically, his theory is that RNA was the primeval parasite.

The French virologist Patrick Forterre ("A hypothesis for the origin of cellular domain", 2006), instead, thinks that today's living beings are descendants of three RNA viruses. These RNA viruses originally evolved the double-stranded DNA molecule to defend their RNA genes, and eventually this "shield" took on a life of its own and became the main mechanism for bacteria, archaea, and eukaryota. It is a fact, that the genes of viruses seem to date back in time to before the birth of cell-based life.

The genetic code is just a code that relates mRNA triples and protein's aminoacids. The genetic code is the same for every being. It is just a code. It translates the instructions in the genotype into a phenotype. But it is an extremely sophisticated code. Did the genetic code itself evolve from a more primitive code? It is unlikely that the first self-replicating organisms were already using today's genetic code. How did the genetic code arise? And why don't we have any evidence of a pre-existing system of replication? Why is it that today there is only one code, rather than a few competing codes (just like there are a few competing genomes)?

Chance

The ultimate meaning of the modern synthesis for the role of humans in nature is open to interpretation. One particular, devastatingly pessimistic, interpretation came from the French biologist Jacques Monod: humans are a mere accident of nature.

To Monod, living beings are characterized by three properties: teleonomy (organisms are endowed with a purpose which is inherent in their structure and determines their behavior); autonomous morphogenesis (the structure of a living organism is due to interactions within the organism itself); and reproductive invariance (the source of information expressed in a living organism is another structurally identical object - it is the information corresponding to its own structure).

A species' teleonomic level is the quantity of information that must be transferred to the next generation in order to assure transmission of the content of reproductive invariance. Invariance precedes teleonomy. Teleonomy is a secondary property stemming from invariance.

All three pose, according to Monod, insurmountable problems. The birth of teleonomic systems is improbable. The development of the metabolic system is a superlative feat. And the origin of the genetic code and its translation mechanism is an even greater riddle. The combined probability is zero. Monod concluded that humans are the product of chance, an accident in the universe.

The paradox of DNA is that a mono-dimensional structure like the genome could specify the function of a three-dimensional structure like the body: the function of a protein is under-specified in the code. Therefore it must be the environment that determines a unique interpretation. There is no causal connection between the syntactic (genetic) information and the semantic (phenotypic) information that results from it.

Then the growth of our body, the spontaneous and autonomous morphogenesis, rests upon the properties of proteins.

Monod concluded that life was born by accident. Then Darwin's natural selection made it evolve, and that process too relied on chance. Biological information is inherently determined by chance.

Life is not the consequence of a plan embodied in the laws of nature: it is a mere accident of chance. It can only be understood existentially. Monod reduces "Necessity", i.e. the laws of nature, to natural selection.

In the 19th century the French physicist Pierre Laplace suggested that, known the position and motion of all the particles in the universe, Physics could predict the evolution of the universe into the future. Laplace formulated the ultimate version of classical determinism: that the behavior of a system depends on the behavior of its parts, and its parts obey the deterministic laws of Physics. Once the initial conditions are known, the whole story of a particle is known. Once all the stories of all the particles are known, the story of the whole system is known. For Laplace, necessity ruled and there was no room for chance. Monod shattered this vision of reality and made it even worse for humans: we are not robots, deterministic products of universal laws, but mere products of chance. In Monod's world, chance plays the role of rationality: chance is the best strategy to play the game of life. Chance is necessary for life to exist and evolve.

Chance alone is the source of all innovation and creation in the biosphere. The biosphere is a unique occurrence non reducible from first principles. DNA is a registry of chance. The universe has no purpose and no meaning.

Monod commented: "Man knows at last that he is alone in the universe's unfeeling immensity out of which he emerged only by chance".

In reality, what Monod highlighted is that the structures and processes on the lower level of an organism do not place any restrictions on higher-level structures and processes. Reality is layered into many levels, and the higher levels are free from determinism from the lower levels. What this means is that high-level processes can be influenced as much from "above" as they are from "below". Monod's "chance" could simply mean "environment" (which even leaves open the possibility of the super-environment of a god influencing all systems).

The German biophysicist Bernd-Olaf Kuppers thinks that there is nothing special about life: all living phenomena, such as metabolism and inheritance, can be reduced to the interaction of biological macromolecules, i.e. to the laws of Physics and Chemistry. In particular, the living cell originated from the iterative application of the same fundamental rules that preside to all physical and chemical processes. Kuppers favors the hypothesis that the origin of life from inorganic matter is due to emergent processes of self-organization and evolution of macromolecules. But, in the balance between law and chance, only the general direction of evolution is determined by natural law: the detailed path is mainly determined by chance. Natural law entails biological structures, but does not specify which biological structures.

To contrast Monod's existential pessimism, Freeman Dyson wrote: "The more I examine the universe and study the details of its architecture, the more evidence I find that the universe in some sense must have known that we were coming."

Necessity

If Monod thought that life was highly improbable and happened only by chance, the US biologist Harold Morowitz believes that life occurred so early in the history of the planet because it was highly probable.

Based on the chemistry of living matter, Morowitz argued that the simplest living cell that can exhibit growth and replication must be a "bilayer vesicle" made of "amphiphiles" (a class of molecules, that includes, for example, fatty acids). Such a vesicle, thermodynamically speaking, represents a "local minimum" of free energy, and that means that it is a structure that is likely to emerge spontaneously. The bilayers spontaneously form closed vesicles. The closure (the membrane) led to the physical and chemical separation of the organism from the environment. This is, for Morowitz, the crucial event in the early evolution of life. Later, these vesicles may have incorporated enzymes as catalysts and all the other machinery of life. These vesicles are the "protocells" from which modern cells evolved.

In other words, Morowitz believes that first came membranes: first membranes arose, then RNA, DNA or proteins or something else originated life. First of all an organism has a border that differentiates it from the environment, that isolates it thermodynamically, that bestows it an identity, that enables metabolism. The second step is to survive: the membrane's content (the cell) must be able to interact with the environment in such a way that it persists. Then the cell can acquire RNA or DNA or whatever else and reproduce and evolve and so forth.

All of this happened not by chance, but because it was very likely to happen. It was written in the laws of Physics and Chemistry.

Furthermore, Manfred Eigen refuted Monod's thesis by showing that natural selection is not blind. Eigen agrees with Monod that information emerges from random fluctuations (from chance), but he thinks that evolution does not act blindly. Evolution is driven by an internal feedback mechanism that searches for the best route to optimal performance.

Eigen found that the distribution of variants is asymmetric, and tends to favor the "best" variants (from a survival point of view). Life seems to know where to look for best variants. As a matter of fact, Eigen discovered a feedback mechanism, inherent in natural selection, that favors (or accelerates the search for) superior variants. Selection is not blind because it is driven by this internal feedback mechanism. Evolution is inherently biased towards the "best" possible solution to the survival problem, and this creates the illusion of the goal-directedness of evolution.

Evolution is "directed" towards optimization of functional efficiency.

Where Monod thinks that (biological) information arises from non-information by sheer luck, Eigen thinks that a fundamental law drives non-information towards information.

The Inevitability of Life

Kauffman proved that life is vastly more probable than traditionally assumed.

The US physicist Jeremy England ("Statistical physics of self-replication", 2013) showed that when matter is driven by a strong external source of energy (like the sun) and surrounded by a "heat bath" (like the sea), it tends to restructure itself in order to dissipate increasingly more energy, i.e. to behave like living matter. That "restructuring" can occur in many ways but two are obvious: self-replication and self-organizing. These are two processes that cause (allow for?) a system to dissipate increasingly more energy.

The Belgian (but Russian-born) physicist Ilya Prigogine had analyzed the behavior of open systems near equilibrium. The behavior of systems that are far from equilibrium because driven by stronger external sources of energy was studied by the Australian physicist Denis Evans ("Probability of Second Law Violations in Shearing Steady States", 1993) and by the Polish physicist Chris Jarzynski ("Nonequilibrium Equality for Free Energy Differences", 1997). Then the British chemist Gavin Crooks ("Entropy production fluctuation theorem and the nonequilibrium work relation for free energy differences", 2008) discovered a simple law: the probability that atoms will undergo a thermodynamic process divided by the probability of the same atoms undergoing the reverse process (such as reconstituting the original lump of sugar) increases as entropy production increases. In other words, the system's behavior becomes more and more irreversible. From these observations England derived his theory.

At the same time others showed that self-replication is not a property of living beings alone. Philip Marcus ("Three-Dimensional Vortices Generated by Self-Replication in Stably Stratified Rotating Shear Flows", 2013) and Michael Brenner ("Self-replicating colloidal clusters", 2013) have discovered it in nonliving matter too.

A History of Life

The Belgian biologist Christian de Duve assembled a detailed explanation of how life started and developed, an explanation that is consistent with the data available from Geology, Paleontology and Anthropology.

One of the guiding principles in his search for the origins of life is that the same principle that gave rise to the chemistry of life ("proto-metabolism") must preside over the chemistry of today's life (metabolism).

Life started, in his opinion, with the spontaneous formation of organic molecules that are widely available in the universe. Organic matter is made of a combination of Carbon, Hydrogen, Nitrogen, Oxygen, Phosphorous and Sulfur (the "CHNOPS" principle). The prebiotic conditions of the Earth enabled them to grow in a recursive relationship that eventually gave rise to nucleic acids and proteins. Life is this network of mutually binding chemical reactions. Life was bound to rise under the conditions of prebiotic Earth.

In sharp contrast with Monod, life turns out to be a deterministic process that is likely to occur whenever the proper conditions are in place.

DeDuve then analyzes how "base pairing" (the "doubling" in the double helix of DNA) is but a special case of a general mechanism of nature, "molecular complementarity". This phenomenon opened the "age of information", in which chemistry that had nothing to do with transmitting information gave rise to replication, inheritance and evolution, processes which are based on information. RNA emerged before proteins did and was responsible for the survival and reproduction of the early forms of life. RNA molecules were the first catalysts of life. Catalysts sped up the chemical reactions required by life. Because of the fragility of proto-life forms, the process that led to RNA molecules must have been extremely rapid.

Replication was initiated by single-stranded RNA molecules but soon led to double-stranded nucleid acids. The mechanism of pairing naturally enables the process of replication, as originally noted by Crick himself (the double works as a negative and a positive, one being the template for the assembly of the other). RNA molecules made of the four A,G,U and C bases had the advantage that could be replicated, thanks to base pairings. RNA genes were born. Selection began operating. Protein synthesis began occurring.

The next quantum leap was the formation of the genetic code and the assembly of a translation apparatus. Then, the separation of replication and translation gave rise to DNA.

Membranes, i.e. outer defenses, were born because the protocell had to devise efficient ways to derive energy from the environment (transmitting signals from the cell to the environment and viceversa, binding with the environment). Life became a property of discrete, autonomous units.

At the same time, cell division began to support replication.

Information-based chemistry allowed for the assembly of a cellular structure, which is the one common ancestor to all forms of life on Earth.

Multi-cellular organisms were created over a long period of time (possibly as long as one billion years). Prokaryotes (bacteria) evolved into eukaryotes: the cell grew more complex, the cell became capable of eating other cells, the cell established "endosymbiosis" (permanent symbiosis) with other cells.

The next accelerating factor was sexual reproduction, again due to constrained chance, which led to the biodiversity we are familiar with in our age and to the complex interplay of organisms within the same ecosystem. The next major step was the development of brains, and the advent of consciousness, which is now reshaping the course of life on Earth. Both life and mind are deterministic consequences of the matter of this universe, not mere chance events.

Each step in the growth of life was providing an incremental selective advantage.

DeDuve believes in one and only one origin of life for the simple reason that life is one: there is only one "life" we are familiar with, the one made of genetic code, metabolism, etc. All "living" creatures share the same "living" processes.

A leitmotiv of the evolution of life is "constrained contingency": mutations occur by chance, but are constrained by physical, chemical and environmental factors.

DeDuve therefore reconciles chance and necessity.

Complexity and Specialization
Darwin himself objected to the idea that there might be a trend towards complexity in nature. Nothing in the laws of evolution implies that life should evolve towards increased complexity. Nevertheless, the facts seem to tell a different story: eukaryotic cells are more complex than prokaryotic ones, animals and plants are more complex than protists, and so on.

The British biologist Ronald Fisher, tried, indirectly, to justify that fact with his fundamental law: the rate of increase in the average fitness of a population equals the genetic variance in fitness. This law is like the second law of Thermodynamics, which implies that entropy can never decrease. Fisher's law says that the average fitness of a population can never decrease (because variance is never a negative number).

The British biologist John Maynard-Smith and the Hungarian biologist Eors Szathmary argued against Fisher's theory and instead proposed that the increase in complexity may originate from very few episodic evolutionary transitions whose goal was not to increase complexity. Their "major transitions" share a common aspect. Each transition affected biological units that were capable of independent replication, and each transition turned them into biological units that needed other biological units in order to replicate. For example, independently replicating nucleid acids evolved into chromosomes (assemblies of molecules that must replicate together). Also, sexless life was replaced by species that have male and female members, and that can replicate only if a male and a female "cooperate". Ants and bees can only replicate in colonies.

Another side of the same coin is the history of specialization. How this happened is not clear but there must have been a point in time when a set of identical organisms "deteriorated" (or, better, differentiated) into functionally specialized organisms. There was a time when only RNA existed; that world decayed into a world of DNA (that carries out the genetic functions) and proteins (that carry out the function of catalysts). The monolithic cells of prokaryotes evolved into the combination of nucleus, cytoplasm and organelles of the eukaryotes. A world of hermaphrodites morphed into a world of sexual organisms. The members of beehives have specific roles. And so forth.

In these major transitions, sets of identical biological units were replaced by sets of specialized units that needed to cooperate in order to survive and replicate.

Maynard-Smith and Szathmary interpret these transitions also on the basis of information theory: they involve a change in the language that encodes information and a change in the medium that expresses that language. In other words, they are about the way in which information is stored and transmitted.

Maynard-Smith defined progress in evolution as an increase in information transmitted from one generation to another.

The key to evolution is heredity: the way information is stored, transmitted and translated. Evolution of life as we know it relies on information transmission. And information transmission depends on replication of structures.

Evolution was somewhat accelerated, and changed in character, by and because of dramatic changes in the nature of biological replicators, or in the way that information is transmitted by biological replicators. New kinds of coding methods made possible new kinds of organisms.

Today, replication is achieved via genes that utilize the genetic code. But this is only the latest episode in a story that started with the most rudimentary replicators. RNA is capable of playing both the roles of replicator and enzyme, as discovered by the US biophysicist Carl Woese. Thus Maynard-Smith thinks likely that the first replicators were made of RNA.

Szathmary showed that this would also explain why the genetic alphabet consists of four letters: four bases are optimal for ribo-organisms. The genetic alphabet evolved when enzymes were ribozymes and organisms with protein enzymes have simply inherited it. At first RNA molecules performed both the job of information management and of constructing the structures specified in that information.

The first major breakthrough in evolution, the first major change in the technique of replication, was the appearance of chromosomes: when one gene is replicated, all are.

A second major change came with the transition from the solitary work of RNA to the dual cooperation of DNA and proteins: it meant the shift from a unitary source of replication to a division of labor: on one hand the nucleic acids that store and transmit information (i.e., the birth of the genetic code as it is today), and on the other hand the proteins that construct the body. Metabolism was born out of that division of labor and was facilitated by the chemical phenomenon of autocatalysis. Autocatalysis allows for self-maintenance, growth and reproduction. Growth is autocatalysis.

Early on, monocellular organisms (prokaryotes) evolved into multicellular organisms (eukaryotes). The new mechanism that arose was gene regulation: the ability to switch on different genes in different cells depending on the stimuli that the cell receives. The code didn't simply provide the instructions to build the organism, but also how cells contributed to the organism.

Asexual cloning was eventually made obsolete by sex, and sex again changed the rules of the game by shuffling the genetic information before transmitting it. The living world split into animals, plants and fungi that have different information-transmission techniques.

Individuals formed colonies, that developed other means of transmitting information, namely "culture"; and finally social behavior led to language, and language is a form of information transmission itself.

Each of these steps "invented" a new way of coding, storing and transmitting information.

Maynard Smith does not continue the story to what is truly unique about humans: morality. Over the centuries humans have progressively abandoned or at least decried old habits such as war, torture, slavery, racism, gender discrimination, pollution.

Maynard-Smith also introduced Game Theory into Biology. The premise of game theory is that individuals are rational and self-interested Maynard Smith applied this definition to populations (instead of individuals) and interpreted the two attributes biologically: rationality means that population dynamics tend towards stability, and self-interest means fitness relative to the environment.

Carbon Chauvinism

Life on Earth uses carbon-based molecules and a base-4 genetic code. Is this part of the definition of life? Is it possible for a living being from another planet to be made of something else and be encoded in a different kind of code, or life is possible only for carbon-based molecules and base-4 genetic codes?

There are simple chemical properties that made carbon-based molecules more efficient for creating the kind of life that prospers on Earth. It is, in fact, relatively easy to prove that no other kind of molecules could provide such an effective medium for the creation of evolving, reproducing and growing bodies.

Nonetheless, it is not clear yet if life "has" to be based on carbon, if non-carbon forms of life are possible.

Humans have built robots made mostly of metal and copper that are capable of reproducing, growing, communicating and so forth, i.e. that satisfy the ordinary definitions of life. This is a very simple example of life that does not use Carbon-based molecules and water. If it is possible on Earth itself, it is hard to believe that non-Carbon life is impossible anywhere in the universe. It is not easy to determine which one is more likely to "spontaneously" arise in nature, an eye or one of these robots. (The "spontaneously" is in quotes because nothing is truly spontaneous: an eye is the product of natural forces just like a robot is, so far, the product of some human design).

Most calculations of the probabilities of carbon-based life are done by scientists who are biased by the fact that they themselves are made of carbon-based molecules. Earthly scientists (made of carbon-based molecules) do not calculate the odds that a robot (made of steel and copper) or some other form of life could emerge in a different kind of planet or star, where, for example, some odd natural phenomena produce stainless steel and copper wires by the millions.

Most Earthly scientists who talk about "another form of life" end up talking about the Earthly form of life (and therefore proving that carbon-based life is the only one possible).

The truth is that is a bit premature to claim that only carbon-based life is possible in this universe.

Also, it is relatively easy to build purely software systems that exhibit whatever property one ascribes to life. These software systems do not use Carbon-based molecules or water: in fact, they use no chemistry at all.

The real issue is that biologists do not agree on a definition of life. If we don't know what life is, it is hard to discuss... what life is.

The Origin of Adaptation
Living organisms exhibit a striking property: their parts and their behavior are adapted to ensure the survival and the reproduction of their entire body. Not only the parts: the behavior too. Animals are born knowing what to do to survive.

Adaptation is a fact, not an opinion. How it came to be is an opinion, not yet a fact.

According to Jean-Baptiste Lamarck, the French botanist of the 19th century who had already claimed in 1802 that animal species are not immutable, acquired characters are inherited: each generation passes on to the following generation what it has learned about adapting to the environment. So far, evidence is against Lamarck: genes do affect proteins, but proteins cannot affect genes. It is a one-way process, from genes to bodies. Once they have been created, bodies cannot change their genes. No matter how much they learn, bodies cannot store it in their genes and pass it on to future generations.

The evidence against Lamarckism is overwhelming. Every generation has to re-learn what previous generations learned. After thousands of years of civilization, children are still born unable to write and to count. Worse: after millions of years, we are still born unable to walk and to speak. According to Lamarck, humans should have already manufactured genes about walking upright and speaking.

We have not found any evidence that a body can purposely alter its own DNA or the DNA it will pass to the offspring. DNA changes only because of random errors in copying. The DNA of a species is manufactured over millions of years by natural selection: the errors that survive become permanent instructions for future generations. But each individual is stuck with the DNA it receives at birth.

Even if he was wrong about the specific mechanism for evolution, Lamarck had powerful insights in the way Nature works on a large scale. In particular, he argued that all of Nature reflects a few general organizing principles. Foremost among them is the effect of use and disuse of organs: muscles atrophy if they are not exercised and bones grow stronger at points where muscles are attached and produce tension.

What Darwin proposed was not "the" theory of evolution (which had already been proposed by many thinkers, including Lamarck himself), but a particular mechanism for evolution: the differential rate of reproduction, under pressure from the environment, of different sorts of individuals within a population; i.e., the differential survival and reproductive success of units of different adaptive efficiency. The key point of Darwin's theory is that variation and selection are dual aspects of the same problem. Lamarck proposed instead a transformational (rather than variational) mechanism.

Later, Darwin's theory of evolution by selection of that variation was indirectly supported by Mendel's mechanism for the inheritance of variation.

However, the US psychologist James-Mark Baldwin discovered what is now known as the "Baldwin effect" ("A New Factor in Evolution", 1896): the ability of individuals to learn can guide the evolutionary process, i.e. the ability to learn can affect evolution. Baldwin was interested in the long-term evolutionary effects of environmental changes. For example, organisms that move to a new ecological niche indirectly subject their descendants to selection pressures which are different from the ones experienced by their ancestors; i.e., the selection pressures that generated their own generation are different from the ones that will generate future generations. It is therefore possible that future generations will evolve because of the change in selection pressure due to the new environmental conditions. The ancestors had to "learn" how to behave in the new environment, whereas the descendents will behave by instinct in that same environment. Thus learned behaviors may become instinctive behavior in subsequent generations, without requiring Lamarck's inheritance. He proved that evolution under those effects is more rapid than in a situation of no change.

In 1958 the Austrian physicist Erwin Schroedinger said something similar: behavior can indirectly alter genetic code, by enabling organisms to survive and reproduce where non-intelligent organisms would simply die.

The British geneticist Conrad Waddington ("Genetic Assimilation Of An Acquired Character", 1953) discovered "genetic assimilation", the process of differential selection by which an individual's response to an environmental stimulus can eventually become a fixed behavior in the species even in the absence of stimulus. Waddington also pointed out that the behavior of a living being changes the environment, and therefore helps to create the selection pressure that will influence its own evolution. Both phenomena point to the importance of the behavior of the organism for its own evolution in a sort of Lamarckian fashion.

Likewise, the German zoologist Ernst Mayr argued that any change in behavior by a population (for example, the acquisition of a new habit due to a move to a new ecological niche) has an effect on the selection pressures that will operate on that population.

The Origin of Speciation

The modern synthesis offers a powerful paradigm for the evolution of life, but looks still inadequate to explain the origin of species. The problem is that it is very difficult to create a new animal. New traits must be assembled in such a way that they allow the organism to survive at least a few generations. The new traits must stabilize. The mutating individuals must avoid being rapidly re-assimilated into the original species through interbreeding. And, last but not least, both sexes must arise at the same time.

Darwin did not solve the mystery of the origin of species, despite the fact that he titled his book that way. Only after the advent of genetics, and mainly thanks to the work of the German zoologist Ernst Mayr ("Change of genetic environment and evolution", 1954), were biologists able to advance hypotheses. If a population splits in two because of whatever accident, both random mutations and environmental differences (natural selection) will cause the two groups to evolve differently until they become two separate species. If the two groups ever meet again, they are more likely to compete than to interbreed, as any species that have similar behavior in the same territory.

Today, there is consensus that species are born (at least) from geographic isolation of a population, but then it is not clear what biological mechanism originates hybrid infertility (this population cannot inbreed with other populations anymore) and fertile diploids (this population has both male and female that can breed).

It is not clear whether the same sudden discontinuity at the level of the phenotype (the organism) occurs as well at the level of the genotype (its genome). We know that the genotype is not the same for every member of a species, that small changes occur all the time. It may well be that changes can accumulate for a long period of time without any visible consequence on the phenotype while they are reaching a crisis point. At that point of genetic "drift", the smallest change in the genotype may have catastrophic consequences for the phenotype.

In the opinion of Harold Morowitz, this is the way organisms acquire new levels of organization, the way they evolve towards more and more complexity. All of a sudden a "gateway" opens up that leads to a whole new range of possibilities. For example, a glue that can hold cells together may be responsible for the sudden appearance of multicellular organisms which in turn quickly acquired a completely new behavior.

The US linguist Philip Lieberman believes in "functional branch-points". He recalls two principles. The first principle is that natural selection acts on individuals who each vary: species that successfully change and adapt are able to maintain a stock of varied traits coded in the genes of the individuals who make up their population. The second principle is the "mosaic" principle, which holds that parts of the body of an organism are governed by independent genes. There are no central genes that control the overall assembly of the body. Given these principles, a series of small, gradual changes in structure can lead to an abrupt change in the behavior of the organism; and an abrupt change in behavior may cause an abrupt change in morphology which causes the formation of a new species. New species are formed at "functional branch-points".

By surveying "adaptive radiation" (the spread of species of common ancestry into different niches) and "evolutionary convergence" (the occupation of the same niche by outcomes of different adaptive radiation), the US biologist Edward-Osborne Wilson argued that opportunity is likely to cause an explosion of species formation.

The problem is that the genetic mechanism that fosters variation is not well understood. Several researchers have observed that bacterial cells tend to choose for themselves advantageous mutations over harmful mutations. Darwinism and modern genetics, instead, prescribe that mutations must be absolutely random. One possible explanation that is not in conflict with Darwinism is that, under conditions of stress, cells generate many more mutations than they would normally do, and of these mutations the most advantageous survive and are observed. If confirmed, this would imply that cells know when the survival of their species is in jeopardy and enter a state of frenzy in which they produce as many mutations as possible, hoping that at least one will be able to adapt and resolve the stress. Natural selection is certainly a weak process for evolving species, but it would be far more effective if it turns out that it is coupled with another process which is capable of generating a lot of diversity every time evolution is desirable because of environmental pressure.

Design Without Progress
A not so subtle argument has to do with the concept itself of "evolution". Evolution intuitively implies a progress from less to more, from lower to higher. Whether Darwin intended it that way or not, the idea that species evolve towards better and better beings does not follow logically from his premises. In particular, any change in the genes is more likely to do harm than to do good to the organism: how can this possibly lead to better organisms?

The US paleontologist Stephen Jay Gould does not believe that there is any inherent "progress" towards bigger complexity in evolution. Life evolves largely by accident. He opposes a biased interpretation of the fossil record. For example, he pointed out that bacteria still represent the dominant form of life on this planet. One should focus on variety and diversity, not "complexity". He objected to choosing one feature as representing a trend. If one considers the whole diversity of life, there is no trend towards progress or higher complexity. Simple forms still predominate in most environments.

We are unlikely accidents, not the fruit of progress. Any replay of the tape of life would yield a different, unpredictable evolutionary history, albeit still a meaningful one. Evolution is not in the hands of determinism and not in the hands of randomness, but in the hands of contingency. Chances that humans would be recreated if history were played back are kind of slim. Gould thought that consciousness evolved only once in all the experiments that life performed on Earth (whereas eyes evolved many times in many species, and so did wings, in both birds and insects). Consciousness is therefore unlikely to occur, and human consciousness must be considered a sheer accident. If the tape of life were played back again, it is unlikely that a conscious being would emerge.

Ernst Mayr put it bluntly when he argued that evolution does not seem to reward smart organisms over stupid ones.

On the other hand life may be more probable than it appears to be, since it happened on Earth as soon as it could happen.

As Francis Crick put it, natural selection has the function of making unlikely events very common.

The mind itself came into the picture quite late in the evolutionary process. If mind is unique to humans, then a tiny change in the evolutionary chain could have resulted in no humans, and therefore no mind. Mind does not look like a momentous episode, but as a mere accident.

Evolution is still a blind process. At any point in evolutionary history the outcome is uncertain. Evolution does not proceed towards complexity but randomly produces variety. Progress is purely accidental. If we interpret Darwin literally, there is only variation, not progress.

Evolutionary Crises

Francis Galton, Darwin's cousin, did not believe that species could be created "gradually", as Darwin's theory implied. Galton believed that new species can arise only in bursts, and so he modified Darwin's theory in a key way. He reasoned that variation allows for a lot of change, but only within a species. A species is a stable state, and has a lot of resilience. When something destabilizes that state, then there occurs a sudden change, and a new species is created, i.e. a new stable state is rapidly reached.

The weakness of Darwin's theory to explain the complexity of life led the German geneticist Richard Goldschmidt in the 1930s to similarly conclude that evolution must proceed by great leaps rather than by small steps.

Expanding on Galton's theory, Gould introduced the idea of "punctuated equilibrium" ("Punctuated equilibria", 1972): evolution occurs through rapid bursts of speciation after long periods of stasis, as opposed to the traditional view of gradual, continuous unfolding of species. Mostly nothing happens; however, when something happens, it happens quickly. Incidentally, we know for sure that this is the way progress occurred in human civilization: long periods of stasis were followed by sudden bursts of progress (we are living in one of those).

About 600 million years ago the first living beings that exhibited bilateral symmetry appeared on Earth. These "bilaterians" were symmetrical beings, characterized by the mirror-image balance of most limbs and organs (the notable exception being the heart, which is still asymmetric to this day). Today bilaterians rule the planet. Bilaterians include most species, from insects to mammals: two legs, two wings, two eyes, two lungs, two brain hemispheres, two kidneys, two nostrils, etc. Organs and limbs that are not duplicated (such as the mouth or the anus or the penis) are located exactly along the axis of symmetry.

It is not clear what the evolutionary advantage of bilaterians was. Although many possible candidates can be advanced, none seems to justify the sudden domination by bilaterians. The early ones were microscopic, and already rather complex. Basically, the rest of evolution was only a magnification of something that had already been created 600 million years ago.

A few million years later there was a sudden explosion of species (the "Cambrian explosion"). Again, several hypotheses have been advanced to explain why suddenly species multiplied rapidly (that animals began to alter the environment, that animals developed emotional responses that made them more likely to survive, that they started eating each other, etc), but none seems to justify the fossil record.

The Evolution of Evolution

There seems to have been waves of evolution, not a linear process of evolution.

The ancestor form of all Earthly life evolved about 3.8 billion years ago and quickly stabilized into the DNA we know today.

Then eukaryotic cells evolved about 2 billion years ago and quickly stabilized into the cell with a nucleus that we know today.

Then multicellular organisms evolved about 540 million years ago (the "Cambrian explosion") and quickly stabilized.

Each wave of evolution created a new degree of complexity in life. It was followed by a long period of relative stability. Then suddenly another "explosion" led to a higher level of evolution.

Design With a Designer

The complexity of life, as we observe it today, is such that the Darwinian mechanism alone may not be enough to explain how it evolved (in fact, Darwin himself thought so, and introduced sexual selection next to natural selection). There might be other processes that help species evolve.

The US chemist Michael Behe is skeptical about Darwin's theory of evolution because cells are too complex to have evolved spontaneously. Most cellular systems are "irreducibly complex", i.e. they could not work without some of their parts. If one of the parts is not there, the system does not operate, and therefore cannot reproduce and evolve.

Such systems cannot be built by gradual evolution: too many of their parts must be there in order for them to be able to start evolving. Their structure cannot be due to evolution because their function cannot be built incrementally. For example, a mousetrap is a mousetrap as long as it has a spring: a mousetrap with the spring cannot evolve from a mousetrap without a spring because the latter would have no function, therefore would simply not survive.

Organisms are even more complex than mousetraps: they require sophisticated mechanisms for storing and transporting enzymes and proteins, among other things. The cell is too complicated, and it needs to be that complicated in order to be a living cell, and therefore it cannot have evolved from something that was less complicated.

Ultimately, Behe's argument is that intermediate stages of evolution do not have intermediate payoffs, and the evolution towards a self-sufficient system with an evolutionary advantage is a sequence of such intermediate stages. The chances that life survives so many intermediate stages are basically zero.

Behe reformulates the main objection against Darwinism: how is it possible that complex organisms "evolved", if their parts must be present all at the same time for the organism to work properly? An eye without the retina is not much use. Even an eye with a retina but without the proper connection to the brain is not much use. All these things have to evolve at the same time in order for an organism to be able to see, or eat, or walk. It is hard to believe that by chance alone something as complex as a living organism would be created.

Behe believes that life must have been designed by an intelligent agent, but, of course, there may simply be other laws at work besides the basic Darwinian laws.

An eye cannot arise suddenly from an eye at all (the odds are on in billions). An eye may arise from something slightly different from an eye. And this "something" may arise from something slightly different from itself, and so forth. This sounds a whole lot more plausible, but: 1. At each step of the "arising" a stable system must be generated ("stable" as in "capable of surviving and reproducing"); and 2. That system must survive long enough to reproduce, otherwise there would be no further evolution. In order for "homo sapiens" to arise, billions of small evolutions must have occurred in all of our organs. The odds are indeed very low. It is debatable whether there has been enough time (i.e., if two billions of years are enough) for these very unlikely events to have occurred. As a matter of fact, a number of biologists began searching for the "accelerator" that may explain how evolution of such complex organisms can occur in such a "short" time and with such efficiency (again, Darwin himself had done so when he introduced sexual selection).

The US biologist Howard Pattee was skeptic too, based on the simple observation that Darwin's blind variation in a virtually infinite search space is inadequate to explain the amazing rate of success at creating species that actually survive.

The British biologist Richard Dawkins answers Behe's objection with a simple argument: even the most sophisticated organ is far from being perfect. We perceive only a fraction of the world. Our eyes don't see and our ears don't hear most of what is out there. Even the animals with the sharpest senses miss some frequencies. We are easily fooled by a sound or a picture. In other words, we "think" that we are such admirable beings, but the truth is that we are precisely one of those imperfect, partial realizations that Behe views as unlikely.

Dawkins points out that numerous "innovations" were forgotten by nature just because they did not get transmitted. Our organs, far from being the best possible of each kind, are merely (my definition) "what survived of what arose".

Post-Darwinian Evolution

Darwin provided a general paradigm, but hardly the final answer. The reason that he can be so easily attacked is that his theory is not really a scientific theory.

A scientific theory provides formulas that one can use to check if the theory's predictions correspond with the behavior of Nature. Darwin's theory of evolution does not provide any formulas. And it would take millions of years to watch a (large) species being created.

Things like knees or eyes cannot have been created by Darwinian evolution alone. The number of nerves and muscles and ligaments and veins that must come together in the exact place at the same time is just too high. Just evolving one nerve is an amazing feat of nature. Imagine evolving the very complex structure of a knee. No matter how many millions of years you have, the chances are lower than the chances that Nature builds a skyscraper.

We can prove the existence of a black hole because we can prove all the physics of the black hole here on Earth. And we can detect the motion of objects around the black hole. But we cannot prove that eyes and knees grew out of variation and natural selection. We have never seen a new limb develop in any species. And nobody has invented a machine yet that can simulate it (precisely because we do not know how it happens).

We do not know the formulas that would create such limbs. We do know the (presumed) formulas of the black hole very well. So we can check if the theory of black holes is correct.

The fossil record only proves that different species have existed at different times. Thus it is easy to claim that there was an "evolution" from microbes to humans. But Darwin added an important point: he also claimed that one species descends from another, driven by variation and selection (that's "Darwinian evolution", not just "evolution"). Alas, Darwin did not explain how this would happen.

Since we don't know how to create a species (it implies creating both a male and a female at the same time that cannot have children with others but only between themselves), we cannot prove that Darwin was right. Worse: we cannot prove that he was wrong.

His theory is perfectly reasonable but there are no formulas that we can experiment with. Thus we cannot prove it or disprove it.

If we had a theory on the origin of species, we could create a new species. Instead we have no clue how to produce one.

At the genetic level the anti-Darwinian argument becomes actually stronger. We observe DNA mutations all the time, but they do not result in new species. DNA mutations result in the same species: your DNA mutates all the time, but you are still a human, and your children will still be humans. According to Darwinists, at some point all these cumulative mutations should suddenly lead to a male and a female that constitute a new species; and survive. But nobody has offered a credible explanation (yet) of why and how this would happen.

If we consider each DNA mutation as an experiment, then the anti-Darwinists are proving their theory every single second: your DNA is mutating every single second but you remain the same species.

However, Darwin discovered something that was indeed scientific: variation plus selection may lead to the creation of order (and, he added, this explains evolution). That part ("Variation plus Selection equals Order") was the beginning of the science of self-organization. And that science might turn out to be the right and definitive way to come up with a theory of evolution that also explains what Darwin could not explain: the origins of species.

Beyond Selection

The opposite kind of anti-Darwinist critique was advanced by John Cairns who discovered "adaptive evolution" ("The origin of mutants", 1988): some bacteria can mutate very quickly, way too quickly for Darwin's theory to be true. If all genes mutated at that pace, they would mostly produce mutations that cannot survive. What drives evolution is natural selection, which prunes each generation of mutations. But natural selection does not have the time to operate on the very rapid mutations of these bacteria. There must be another force at work that "selects" only the mutations that are useful for survival. Cairns speculated that bacteria must be somehow able to choose which mutations to undergo in order to adapt to challenging environments. The idea of evolution driven by "directed mass mutations" had been pioneered in the 1920s by the Russian geographer Lev Berg, a fierce anti-Darwinian.

Design as the Designer

The US biologist Stuart Kauffman even hinted that Darwin may have gotten it all wrong: it is not that natural selection creates order, it is order that succeeds despite natural selection.

Kauffman's mathematical model shows that arrays of interacting genes do not evolve randomly but converge toward a relatively small number of patterns, or "attractors". This ordering principle may have played a larger role than did natural selection in guiding the evolution of life.

Order arises spontaneously, it is inherent in nature, and natural selection is a distraction. Systems tend to self-organize, despite any obstacle they may encounter in their struggle for self-organization. Life is but one such system, and its evolution is a result of that property of our universe: that complex systems tend to self-organize.

Natural selection is not the only source of order and design. There is also order and design for free.

The Origin of Variation

The debate on natural selection is mirrored by an equivalent debate on variation. It is not clear how genetic mutations (mutations in the DNA of an animal) result in phenotypic change (variations in the "body" of an animal). The issue, ultimately, relates to the problem of how a limited number of genes and a limited number of gene mutations can give rise to the very sophisticated complexity of living organisms. It also relates to the fact that "change" to a phenotype would seem more likely to harm than benefit the organism. Genetic mutation must be somehow directed for the odds of producing non-lethal phenotypic variation to be reasonable. Randomness alone would be too dangerous. It appears that evolutionary change would just not be feasible without that guiding hand.

The US biologists Marc Kirschner and John Gerhart proposed that variation is "facilitated" by "conserved core processes" that are not subject to change and are shared by all living organisms. These are the functions that allow the organism to survive phenotypic change. They are linked in a loose regulatory network, and they can reconfigure themselves to accommodate changes in non-core processes. For example, the same genes can yield a hand or an eye depending on which one and when and where gets "expressed". Those are factors that the core processes can alter by reconfiguring themselves. These core processes work as an insurance that random genetic mutations be channeled in the direction of phenotypic changes that are, if not useful, at least not harmful to the organism. Indirectly, they "facilitate" change. These processes reduce the requirement for simultaneity in the evolution of novelty.

The theory of facilitated variation still does not explain its own premise: how did facilitated variation arise in the first place among living organisms.

The Israeli biologist Merav Parter ("Facilitated Variation", 2008) proposed that facilitated variation is a natural occurrence in environments that change in a systematic way, by maintaining the same set of subgoals although organized in different combinations. In other words, the microscopic evolution of core biological processes would recapitulate the macroscopic evolution of the environment.

Facilitated variation might help explain "mosaic" evolution, a concept introduced by the British biologist Gavin de Beer in 1954 ("Archaeopteryx and evolution") when he noticed that different traits evolve at different rates. He viewed an organism as a mosaic whose pieces could evolve separately. This might indeed be a key feature of any evolving system: to consist of parts that are so loosely coupled that each can evolved separately from the others without causing a collapse of the system as a whole.

Teleological Evolution
From the beginning Darwin was criticized for his idea that animals are simply objects in the hands of nature's random will.

The British biologist Samuel Butler, a contemporary and fierce enemy of Darwin, countered Darwin's mechanical, Newtonian view of evolutionary laws operating on inert living matter with the idea that life, far from being inert, has "free will" and has used it to influence its own evolution. It is not only humans who can affect their environment to direct their own evolution: the whole environment is doing the same. Living beings make decisions all the time, which, no matter how small, have an impact on the environment, and are thus responsible in part for their own evolution. Life is not inert, powerless in the hands of evolutionary laws similar to Newtonian laws of gravitation: life contributes to shape the environment in which those laws apply; or, equivalently, life is part of the laws of its own evolution. Butler believed that each living being is sentient: it is conscious, it has memory and it can learn. Butler believed that all life is "teleological": it pursues a goal. Darwin had neglected the fact that life contributes to its own evolution.

Butler realized that living beings are even capable of memorizing their behavior as species. A striking fact gives phylogeny (the development of a species) a kind of supremacy over ontogeny (the growth of the individual): the difference between two newborn infants of two different generations is minimal, whereas the difference between the same individual as a newborn and an adult is much greater. Living beings memorize their behavior at the phylogenetic level. You are more similar to me at the same age than to the way you were twenty years ago.

Butler was convinced that the phylogenetic memory was crucial to evolution. He thought that every living being is conscious of doing something the first time, but repeated performance leads to unconscious habits (just like we drive a car without any conscious effort after years of driving) and, more importantly, that unconscious habits eventually find their way into the physiology of the species (an idea akin to Lamarck's acquired character).

Butler believed that evolution was to a large extent controlled by its object, whereas Darwin believed that evolution was a blind force (and even random) operating on its object.

Darwin believed in design without a designer, Butler believed in design through distributed goal-driven behavior, through (indirect) cooperation. Butler believed in the free will of life.

Life and Mind
One wonders if the relationship between life and mind could be turned upside down.

Life is a cycle of chemical reactions that eventually got enveloped in a "skin" and embodied in a body.

Mind (as the set of cognitive faculties, ranging from vision to learning) is, ultimately, a cycle of electrochemical reactions. What if it arose independently but eventually evolved within bodies?

Mind may not require life, but it would probably not have evolved to our stage without a living body to use and serve.

Mind (conceived as this abstract electrochemical phenomenon) may have existed before life. Life gave it a body, a brain. Life gave it a... life.

Mind may not be the only process that was "absorbed" by life. Digestion may have pre-dated life, as a phenomenon freely occurring on Earth. And so many other chemical processes.

Mind is one of the many processes that a body uses to grow, survive and reproduce.

Further Reading
Behe, Michael: DARWIN'S BLACK BOX (Free Press, 1996)
Berg, Lev: NOMOGENESIS (1922)
Butler, Samuel: EVOLUTION OLD AND NEW (1879)
Cairns-Smith, Graham: GENETIC TAKEOVER (Cambridge University Press, 1982)
Calvin, Melvin: CHEMICAL EVOLUTION (Clarendon, 1969)
Cavalli-Sforza, Luigi: GENES, PEOPLES AND LANGUAGES (North Point, 2000)
Crick, Francis: LIFE ITSELF (Simon & Schuster, 1981)
Crick, Francis: ASTONISHING HYPOTHESIS (MacMillan, 1993)
Cziko, Gary: WITHOUT MIRACLES (MIT Press, 1995)
Darwin, Charles: ON THE ORIGIN OF SPECIES BY MEANS OF NATURAL SELECTION (1859)
Davies, Paul: THE FIFTH MIRACLE (Simon & Schuster, 1998)
Dawkins, Richard: THE BLIND WATCHMAKER (Norton, 1987)
Dawkins, Richard: RIVER OUT OF EDEN (Basic, 1995)
Dawkins, Richard: CLIMBING MOUNT IMPROBABLE (Norton, 1996)
Delsemme, Armand & DeDuve Christian: OUR COSMIC ORIGINS (2001)

Dennett, Daniel: DARWIN'S DANGEROUS IDEA (Simon & Schuster, 1995)
DeDuve, Christian: VITAL DUST (Basic, 1995)
Depew, David & Weber Bruce: DARWINISM EVOLVING (MIT Press, 1994)
Deutsch David: THE BEGINNING OF INFINITY (Viking, 2011)
Dobzhansky, Theodosius: GENETICS AND THE ORIGIN OF SPECIES (1937)
Dyson, Freeman: ORIGINS OF LIFE (Cambridge Univ Press, 1985)
Edelman, Gerald: TOPOBIOLOGY (Basic, 1988)
Eigen, Manfred: STEPS TOWARDS LIFE (Oxford University Press, 1992)
Fisher, Ronald Aylmer: THE GENETICAL THEORY OF NATURAL SELECTION (Dover, 1929)
Goldschmidt, Richard: THE MATERIAL BASIS OF EVOLUTION (1940)
Gould, Stephen Jay: ONTOGENY AND PHYLOGENY (Harvard University Press, 1977)
Gould, Stephen Jay: EVER SINCE DARWIN (Deutsch, 1978)
Gould, Stephen Jay: WONDERFUL LIFE (Norton, 1989)
Gould, Stephen Jay: FULL HOUSE (Random House, 1996)
Gould, Stephen Jay: THE STRUCTURES OF EVOLUTIONARY THEORY (Harvard Univ Press, 2002)
Jones, Steven: LANGUAGE OF GENES (Harper Collins, 1993)
Kauffman, Stuart: THE ORIGINS OF ORDER (Oxford University Press, 1993)
Kirschner, Marc and Gerhart, John: "The Plausibility of Life" (2005)
Kuppers, Bernd-Olaf: INFORMATION AND THE ORIGIN OF LIFE (MIT Press, 1990)
Lamarck, Jean-Baptiste: PHILOSOPHIE ZOOLOGIQUE (1809)
Lane, Nick: THE VITAL QUESTION (Profile, 2015)
Lieberman, Philip: UNIQUELY HUMAN (Harvard Univ Press, 1992)
Lorenz, Konrad: "Evolution and Modification of Behavior" (1965)
Mason, Stephen: CHEMICAL EVOLUTION (Clarendon Press, 1991)
Maynard-Smith, John: THEORY OF EVOLUTION (Cambridge University Press, 1975)
Maynard-Smith, John & Szathmary Eors: THE ORIGINS OF LIFE (Oxford University Press, 1999)
Maynard-Smith, John & Szathmary Eors: THE MAJOR TRANSITIONS IN EVOLUTION (W. H. Freeman, 1995)
Mayr, Ernst: ANIMAL SPECIES AND EVOLUTION (Harvard Univ Press, 1963)
Mayr, Ernst: SYSTEMATICS AND THE ORIGIN OF SPECIES (Columbia Univ Press, 1942)

Mayr, Ernst: POPULATION, SPECIES AND EVOLUTION (Harvard Univ Press, 1970)

Mayr, Ernst: THE GROWTH OF BIOLOGICAL THOUGHT (Harvard Univ Press, 1982)

Monod, Jacques: LE HASARD ET LA NECESSITE' (1971)

Morowitz, Harold: BEGINNINGS OF CELLULAR LIFE (Yale University Press, 1992)

Waddington, Conrad: THE STRATEGY OF GENES (1957)

THE PHYSICS OF LIFE

Life has three dimensions. One is the evolutionary dimension: living organisms evolve over time. Two is the reproduction dimension: living organisms are capable of reproducing. Three is the metabolic dimension: living organisms change shape during their life.

Each dimension can be studied with the mathematical tools that Physics has traditionally employed to study matter. But it is apparent that traditional Physics cannot explain life. Life exhibits properties that rewrite Physics.

The Origin Of Self-organization: Life As Negative Entropy

The paradox underlying natural selection (from the point of view of physicists) is that on one hand it proceeds in a blind and purpose-less way and on the other hand produces the illusion of more and more complex design. This continuous increase in information (i.e., the spontaneous emergence of order) seems to violate the second law of Thermodynamics, the law of entropy.

Ludwig von Bertalanffy borrowed the term "anamorphosis" from the German biologist Richard Woltereck to describe the natural trend towards emergent forms of increasing complexity.

Entropy is a measure of disorder and it can only increase, according to the second law of Thermodynamics. Information moves in the opposite direction.

Most things in this universe, if left alone, simply decay and disintegrate. Biological systems, instead, appear from nowhere, then organize themselves, and even grow!

This leads to the "two arrows of time": the behavior of inanimate matter pointing towards entropy increase and therefore disorder increase, and the behavior of biological systems pointing the other way by building increasingly complex structures of order.

When you drop a sugar cube in your coffee, it dissolves: while no physical law forbids the re-composition of the sugar cube, in practice it never occurs, and we intuitively know that it cannot occur. Order is destroyed and cannot be recreated. That's a manifestation of the second law of Thermodynamics. On the other hand, a teenager develops into an adult, and, while no biological law forbids it, and as much as they would like to, adults never regress to youth. This is a manifestation of the opposite arrow of time: order is created and cannot be undone.

Since organisms are made of chemicals, there is no reason why living systems should behave differently than inanimate systems. This is a paradox that puzzled not only biologists, but physicists too.

The German physicist Ludwig von Boltzmann was possibly the first scientist to realize the importance of entropy for life. He reasoned that there is plenty of energy on Earth (air, water, minerals). Life is not driven by energy, or there would be no need for competition: life is driven by competition for entropy. Entropy (created by the transfer of energy from a hot Sun to the cold Earth) is much scarcer.

In the 1940s the Austrian physicist Erwin Schroedinger, one of the founders of Quantum Mechanics, first proposed the idea that biological organization is created and maintained at the expense of thermodynamic order. Life displays two fundamental processes: creating order from order (the progeny exhibits the same order as the parent) and creating order from disorder (as every living system does at every metabolic step, eating and growing). When they create order, living systems seem to defy the second law of Thermodynamics. In reality, they live in a world of energy flux that does not conform to the closed-world assumptions of Thermodynamics. An organism stays alive in its highly organized state by absorbing energy from the environment and processing it to produce a lower entropy state within itself. "Living organisms feed upon negative entropy": they attract "negative entropy" in order to compensate for the entropy increase they create by living. Living organisms can be viewed as swimming upstream against a tide of rising entropy. One can recast life as a cosmic struggle: the existence of a living organism depends on increasing the entropy of the rest of the universe.

In 1974 the Hungarian biologist Albert Szent-Gyorgyi proposed to replace "negentropy" with the positive term "syntropy", so as to represent the "innate drive in living matter to perfect itself". This has a correspondent on the psychological level, "a drive towards synthesis, towards growth, towards wholeness and self-perfection".

Life as Non-equilibrium
In the 1960s the Belgian (but Russian-born) physicist Ilya Prigogine had a fundamental intuition: living organisms function as "dissipative structures" (a term first introduced by the Ukrainian chemist Alfred Lotka). These are structures that form as patterns in the energy flow and that have the capacity for self-organization in the face of environmental fluctuations. In other words, they maintain their structure by continuously dissipating energy. Such dissipative structures reside permanently in states of non-equilibrium, unlike inanimate matter.

Life maintains itself far from equilibrium: the form stays pretty much the same, while the material is constantly being replaced by new material, part of which comes from matter (food, air, water) and part of which comes from energy (sun). The flow of matter and energy "through" the body of the living organisms is what makes it possible for the organism to maintain a (relatively) stable form. In order to stay alive, they have to be always in this state far from equilibrium.

Equilibrium is death, non-equilibrium is life.

And here is the solution of the riddle. Equilibrium is the state of maximum entropy: uniform temperature and maximum disorder. A system that is not in equilibrium exhibits a variation of entropy which is the sum of the variations of entropy due to the internal source of entropy (which tends to increase towards equilibrium) plus the variation of entropy due to the interaction with the external world. The former is positive, but the latter can equally be negative. Therefore total entropy can decrease.

An organism "lives" because it absorbs energy from the external world and processes it to generate an internal state of lower entropy. An organism "lives" as long as it can avoid falling in the equilibrium state.

(In a sense, organisms die because this process is not perfect: if our bodies could be made to keep their shape exactly the same, they would always remain far from the equilibrium and they would never die).

(But then there is a reason why it is not perfect and we have to die: a stable immutable form of life would have scant chances of surviving the continuous changes in the environment, whereas a form of life that continuously reshapes itself has a chance to "evolve" with the environment).

Thanks to the advent of non-equilibrium Thermodynamics, it is now possible to bridge Thermodynamics and evolutionary Biology. By focusing on entropy, structure and information, it is now possible to shed some light on the relationship between cosmological evolution and biological evolution. Biological phenomena can be viewed as governed by laws that are purely physical. This step might prove as powerful as the synthetic theory of evolution.

Prigogine's non-equilibrium approach to evolution, i.e. that biological systems (from bacteria to entire ecological systems) are non-equilibrium systems, has become a powerful paradigm to study life in the context of Physics. Life can finally be reduced to a natural phenomenon just like electromagnetism and gravity.

The Austrian physicist Erich Jansch has extended Prigogine's vision of life to the entire universe: the universe as a gigantic self-organizing system subject to the laws of non-equilibrium thermodynamics.

The US chemist Jeffrey Wicken went as far as to state that "Thermodynamics is above all the science of spontaneous processes", and link life with the expansion of the universe.

Cycles

Lotka pioneered the view of biological systems as endless cycles. Not only is a biological system a network of chemical agents (one chemical reaction leading to another one), but these agents somehow yield a cycling structure. The cycle helps the network exist, assume an identity and grow. This behavior is typical of life. Lotka named it "autocatalysis" ("Contribution to the Theory of Periodic Reactions", 1910).

Indirectly, Lotka was also one of the first scientists to show that biology was chemistry, and chemistry biology.

Bioenergetics

These ideas led to an approach to life, called "Bioenergetics", which consists in applying thermodynamic concepts (energy, temperature, entropy and information) and non-equilibrium (or irreversible) Thermodynamics to biological structures.

The starting point, in the 1920s, was Lotka's assumption that ecosystems are networks of energy flows. In 1941 the German biologist Fritz Lipmann recognized the role of phosphates in biological systems. Then the US brothers Howard and Eugene Odum devised a thermodynamic model for the development of the ecosystem. That became the route followed by an entire branch of Bioenergetics: looking for the thermodynamic principle that guides the development of ecosystems.

In other words, first came the realization that biological systems (living organisms) are about the flow and transduction of energy, i.e. that life is about energy. Then biologists started employing Thermodynamics, the discipline that studies energy. This led to the realization that the Thermodynamics of biological systems is non-equilibrium Thermodynamics, that requires non-linear systems of equations. This led to the development of a new branch of Mathematics that studies non-linear dynamics.

Howard Odum, for example, coined the term "emergy" (for "embodied energy") to refer to the "energy memory" of living systems (a measure of energy used in the past). To him living systems had been formed by an accumulation of past energy, and thus were memories of all that energy.

Eugene Odum viewed the entire Earth as a set of interconnected ecosystems.

The US biologist Harold Morowitz held that the flow of energy through a living system acts to organize the system: organization emerges spontaneously whenever energy flows through a system. The contradiction between the second law of Thermodynamics (the universe tends towards increasing disorder) and biological evolution (life tends towards increasing organization) is only apparent, because Thermodynamics applies to systems that are approaching equilibrium (either adiabatic, i.e. isolated, or isothermal), whereas natural systems are usually subject to flows of energy/matter to or from other systems.

First of all, life is the property of an ecological system, not of a single, individual, isolated organism. An isolated living organism is an oxymoron.

Life of any organism depends on a flow of energy, and, ultimately, life "is" that flow of energy.

Morowitz proved two theorems that analyze what happens during that flow of energy through the chemical systems that living organisms are made of: 1. Those systems store energy in chemical bonds, i.e. their complexity steadily increases; and 2. Those systems undergo chemical cycles of the kind that pervade the biosphere (e.g., the carbon cycle).

The flux of energy turns out to be the organizing factor in a dissipative system. When energy flows in a system from a higher kinetic temperature, the upper energy levels of the system become occupied and take a finite time to decay into thermal modes. During this period energy is stored at a higher free energy than at equilibrium state. Systems of complex structures can store large amounts of energy and achieve a high amount of internal order.

The cyclic nature of dissipative systems allows them to develop stability and structure within themselves.

The bottom line is that a dissipative system develops an internal order. Morowitz proved that Lotka was right: the flow of energy through a (steady state) system yields cycles, which in turn yield structure.

The Evolution of Complex Systems

Jeffrey Wicken placed the second law of Thermodynamics at the center of all biological processes: reproduction, evolution, ecosystems. At all levels of life the second law is not only compatible with life but the very reason for it.

The US biologist Eric Schneider sided with Ilya Prigogine (and his distinguished predecessors Alfred Lotka and Erwin Schroedinger) in thinking that biological systems are the product of non-equilibrium thermodynamics, and indirectly the product of the second law of thermodynamics (that entropy can never decrease, unless there is a flow of energy). Living systems "emerge", or self-organize, and, de facto, it is inevitable that they emerged: it is written in the laws of physics. Non-equilibrium thermodynamics is the science of life. Thermodynamic equilibrium is death.

Schneider believes that there is more than a mere tendency towards creating life. He believes that life is a way to optimize and accelerate (not just carry out) the production of entropy mandated by the second law of Thermodynamics. Living systems are the most efficient way ever devised by Nature to destroy order; and never mind that living systems themselves appear to be among the most ordered systems ever created. It is all a trick to ultimately destroy order, just like warships are sophisticated complex buildings but overall their contribution to history was to destroy buildings.

In 1965, the Polish physicist Joseph Kestin proved that closed systems that are suddenly "freed" (i.e., their constraints are removed) tend to move towards a new state of equilibrium that is an "attractor". This state is called "attractor" because it does not depend on the order in which the constraints are removed: the system "has" to move towards that state. In other words, not only are some processes irreversible, but processes have a direction and an end. This is expressed by his "Unified Principle of Thermodynamics" (It is called "unified" because it really summarizes the other laws of Thermodynamics).

Schneider believes in a natural extension of this principle: a system pushed away from its attractor, will tend to return to the attractor. The stronger the push, the stronger the reaction, the reaction being some form of self-organization. When the gradient pushing the system is particularly strong, the system may self-organize in ever more complex structures.

Thus Schneider believes that nature will create complex systems whenever it can: it will use any means available to achieve equilibrium.

The "actor" that pushes systems away from equilibrium is a gradient (a difference of temperature, pressure or other). Whenever a gradient is applied, the system is no longer in equilibrium. Schneider believes that the reaction to a gradient is internal reorganization aimed at reducing and eventually neutralizing the external gradient. Gradient neutralization is a fundamental property of Thermodynamics, a fact already proven in 1993 by the US physiologist Don Mikulecky.

"Exergy" is the amount of energy that one can extract from a system in the form of work. When the system has reached a condition of equilibrium, its exergy is zero. Exergy is thus also a measure of how far from equilibrium a system is.

An equivalent way of describing this phenomenon is to talk of "gradients", because the gradient is what can be used to generate work: heat, for example, can be turned into work because there is cold, otherwise it would be useless.

Living beings are non-equilibrium systems, so they have high exergy.

Schneider believes that the universal tendency towards equilibrium is driving evolution, that Nature is building more and more complex systems in order to erase exergy ever more efficiently. Living beings are only a cog in the machine built by Nature to destroy all exergy and achieve equilibrium. Life is only a way to break down concentrations of energy and turn it into diffuse waste heat. And we are an accidental by-product of such a universal process. Animals, for example, degrade the exergy of plants when they eat them. Furthermore, they do so in a way that is more efficient than other physical processes (burning the plant, for example, would radiate energy, while a cow eating grass radiates very little energy). In a sense, Nature created animals because they are the most efficient way to erase the exergy of plants, and it created plants because they are the most efficient way to erase the exergy of sunlight, and so on. Ecosystems have evolved from systems that emitted a lot of exergy to systems that emit little exergy. Metabolism is simply a way to degrade energy, and today's animals (such as mammals and birds) are a lot more efficient at it than the first forms of life. Evolution has been progressing towards more and more efficient systems to destroy exergy. Genetic information is simply information about how to destroy exergy.

Ecosystems as a whole can also be viewed as efficient destroyers of gradients, destroyers that are built and driven by energy flows. The thermodynamic study of ecosystems carried out by the British zoologist Evelyn Hutchinson, the Spanish biologist Ramon Margalef, and the US biologist Eugene Odum showed that ecosystems progress towards increased production of entropy and increased reduction of gradient (a different way of saying that biomass, species diversity and energy throughput all increase). The dynamics of ecosystems, in turn, drives evolution. It is the second law that selects the systems that are best at reducing gradients. Natural selection is, in a sense, an afterthought: the most important selection has already occurred, driven by the second law of thermodynamics.

Darwin created a bridge between humans and other forms of life, and explained how one descended from the others. Schneider attempts to do something similar for living systems and non-living systems. Non-living complex systems play the same role that living systems play: they are just a bit less efficient. But they too are driven by the same phenomenon (ultimately, a "gradient"). They too spontaneously organize due to the energy flow caused by the gradient. Thus any non-living complex system appears to be a predecessor of a living system. It is "born" and it "grows" and it "evolves" in a way similar to how life does, except that it is not alive (for example, it does not reproduce).

Schneider summarized his universal principle as "Nature abhors gradients". Whenever there is a gradient, Nature responds by creating the most efficient way to erase it. One such efficient way is life, and us.

In the end, the second law of Thermodynamics turns out to be responsible for both processes of living beings: both their growing and their decaying, both their non-equilibrium (peaking with progress and civilization) and their equilibrium (death).

In the end, the purpose of life turns out to be death: Nature invented life on Earth as the most efficient process to reduce the gradient created by the Sun heating the Earth. The ultimate goal is to reestablish an equilibrium that will, incidentally, destroy all life when life will no longer be needed to reduce a gradient that life will have erased. The meaning of life is, ultimately, suicide.

The Origin of Biological Information
A different but similar non-biological approach to life is based on information, and directly influenced by Cybernetics and Information

Theory. Life is viewed as information capable of replicating and modifying itself.

The pioneering work of the Spanish ecologist Ramon Margalef in the 1960s set the stage. He viewed an ecosystem as a cybernetic system driven by the second law of Thermodynamics. Succession (the process of replacing old species with new species in an ecosystem) is then a self-organizing process, one whereby an element of the system is replaced with a new element so as to store more information at less energetic cost.

The US anthropologist Gregory Bateson argued that the substance of the biological world is, ultimately, "pattern" (not this or that chemical compost), a position that allowed him to seek a unified view of cognitive and biological (and cybernetic) phenomena. His definition of information stretched beyond mere computation: a bit of information "is a difference that makes a difference". Thereby implying that, in order to be information, a pattern must affect something. Nature prefers double/dual systems: two sources of information are infinitely better than one. Hence two eyes and not just one. One eyes would see only in two dimensions. Two eyes reveal a third dimension, depth. Information is the difference that results from the comparison of two things, and it is more than the sum of the parts.

Bateson' thesis was that life is due to dual stochastic processes, each stochastic system driven by a random process and checked by a nonrandom process. One stochastic system is the genetic one. The other one is somatic change. Genetic change is a stochastic process that is triggered by randomness. Somatic change too is a stochastic process triggered by randomness. Both random processes are "selected" by a nonrandom process, whether the need to preserve the organism's internal organization or the need to adapt to the environment. Without the randomness it would be impossible to create something new, there would only be conservation of the old (the nonrandom system). In both of these stochastic processes a large number of alternatives is generated and then the nonrandom selection process prunes that set of alternatives. The combination of the two stochastic processes works in a manner similar to a cybernetic system: a feedback process controls what is actually possible. The source of randomness in one stochastic system is digital (the DNA) and in the other one is analogic (the body and the environment).

The German biophysicist Bernd-Olaf Kuppers found an elegant way to reconcile the paradox of increasing information. Life is biological information, and the origin of life is the origin of biological information. Information has different aspects: syntactic (as in information theory), semantic (function and meaning of information for an organism's survival), and pragmatic (following the German physicist Carl-Friedrich Von Weizsacker, "information is only that which produces information"). Since evolution depends on the semantic aspect of information, there is no contradiction with the second law of Thermodynamics, which only deals with the structural aspect of matter (i.e., the syntactic aspect of information). The origin of syntactic information relates to the prebiotic synthesis of biological macromolecules. The origin of semantic information relates to the self-organization of macromolecules.

The US biologist Christopher Langton emphasized that living organisms use information, besides matter and energy, in order to grow and reproduce. In living systems the manipulation of information prevails over the manipulation of energy. Life depends on a balance of information: too little information is not enough to produce life, too much can actually be too difficult to deal with. Life is due to a reasonable amount of information that can move and be stored. Life happens at the edge of chaos. Ultimately, life is a property of the organization of matter.

As the Canadian biologist Lionel Johnson put it ("The Thermodynamic Origin of Ecosystems", 1998), a bio-system can be compared to an information processor, whose job is to continuously extract, store and transmit information. Two fundamental and opposed forces compete, one leading towards increased uniformity (and lower information) over "ecological" time and one leading towards increased diversity (and greater information) over "evolutionary" time. This results in a hierarchy of living organisms, which has at the top the one species that developed the best strategy of energy extraction and storage, the highest resource utilization and the least dissipation (this is a reworking of a principle due to Alfred Lotka in the 1920s). Extracting information requires an energy flow, which in turns causes production of entropy. This can also be viewed from the point of view of communication: dissipative structures can exist only if there is communication among their components, whether in the form of genetic code (communication over time) or societies (communication over space). The bio-system is, ultimately, an information processor and a communication network.

Predating Freeman Dyson's definition of life, the Hungarian chemist Tibor Ganti views life as the combination of two systems: metabolism and information control. The simplest form of life, in practice, is the "chemoton": an autocatalytic cycle coupled with an information molecule. Ganti's living organism, therefore, looks more like a computer than a program, because it includes the "hardware". Life without the hardware is not life, it is just the process that generates life. It also takes that "information molecule" to have life.

The Primacy Of Energy Flows

The US physicist Ronald Fox argued that the evolution of life is nothing but the evolution of more and more sophisticated forms of transducing and storing energy.

According to Fox, life was born (the transition from non-life to life occurred) thanks to structures for transducing energy that were made possible by the use of phosphate compounds; and that life evolved because these structures inevitably evolved ways to regulate and to store energy via phosphate compounds as a consequence of having employed phosphate compounds in the first place. Muscles and nervous systems arose as a consequence of those phosphate-based processes. Muscles and nervous systems allowed the organism to interact with the environment and with other organisms in a much more intense way, thus further accelerating evolution.

Fox showed how, from the beginning, it was energy flows (lightning, volcanic heat) that allowed for the manufacture of the unlikely molecules (such as aminoacids and other monomers) that are the foundations of life. Monomers had to combine in order to form the polymers of life (notably, proteins), a process that requires polymers. Fox showed that the flow of energy solved this apparent paradox. Energy moved the monomers to a chemical state in which polymerization occurs spontaneously, and then polymers themselves helped manufacture even more polymers. The early Earth's geophysical energy flows gave rise to oxidation-reduction energy. Iron is an obvious source of oxidation-reduction energy, particularly during the "iron catastrophe" of 4.5 billion years ago, when the elements of the iron group percolated through the silicate layer of the Earth's crust and became the core of the Earth. Eventually some oxidation-reduction energy got converted into phosphate-bond energy, which flowed through the primitive organic elements and enabled the polymerization of monomers (notably, proteins, which are both catalysts for further polymerization and structural elements of living beings, and polynucleotides, the building blocks of genetic memory). In other words, energy flows excited monomers until they started creating polymers spontaneously. Eventually, the system reached a state in which polymers helped produce (synthesize) more polymers.

Thus an important form of energy for the evolution of life on Earth was based on phosphorus, which is a relatively rare element.

As the German chemist Fritz Lipmann first observed, the meeting of (pervasive) oxidation-reduction energy and (relatively rare) phosphate-bond energy was a momentous event in the history of life.

Organic molecules are based on carbon, and energy transactions are based on phosphorus. The reason is probably to be found in the chemical properties of these elements. Carbon fosters structural stability, phosphorus fosters energy metabolism.

Life seems to be, ultimately, a process about storing, transducing and using energy. Fox went on to speculate that the very evolution of life (and the diversity of life we observe today on the Earth) was due to an evolution of the processes of energy metabolism and storage. Biological evolution was also driven by energy metabolism and storage, not just Darwinian natural selection.

In other words, Fox argued that, in general, biological events correspond to changes in flows of energy. One key evolution was the emergence of phosphagens as the main tool for energy storage. This more efficient way of storing and using energy enabled the evolution of muscles and motility in general. Phosphagens themselves evolved, and today all higher forms of life employ creatine phosphate.

A key step in the evolution of life was the development of a nervous system. Fox pointed out that the interaction between organism and environment as well as the interaction among organisms are nonlinear in nature. The nervous system is not only capable of predicting the outcome of linear situations, but also of predicting the much more important outcome of nonlinear situations that are, by their own nature, very hard to predict. The reason is that the nervous system allows the organism to rapidly simulate the outcome of nonlinear events. Rapid simulation is the only way that the organism can predict what will happen, and is therefore essential to survival.

A fundamental property of life is the ability to predict the future. Survival depends not so much on being able to calculate what to do, but on being able to predict what is going to happen (in particular, the consequence of an action). A cognitive agent needs to predict situations. Those situations in the real world are described by non-linear systems. The dynamics of nonlinear systems is such that it is virtually impossible to predict their behavior other than by simulating it. Simulation has to be very fast in order to be useful, sometimes faster than real-time. It turns out that the nervous system is precisely such a tool to perform fast simulations of nonlinear systems. Thus the evolutionary advantage of a nervous system is colossal.

Once muscles existed, it was important for the organism to predict the effect of a muscle movement on the environment.

Fox argued that the evolution of creatine phosphate (the ultimate phosphagens so far) enabled the emergence of this wonderful invention. Yet again, the evolution of life was due to an evolution of energy processes.

Darwin's theory of evolution is both too little and too much. Fox showed that the rise and decline of different species is a natural consequence of a non-linear system driven by energy flows. Natural selection is not necessary to explain why there are different forms of life. At the same time natural selection would not be enough to explain how life evolved the way it did.

Fox's theory is all based on the simple idea that whatever happened to life was driven by flows of energy, because ultimately life "is" about storing and using energy. Even culture itself (i.e., human civilization) can be viewed as a new flow of energy that is creating a new form of life.

Fox did more than advance a hypothesis on the transition from non-life to life: he introduced a new factor in Darwin's theory of evolution. The purely Darwinian picture of natural selection may be too little in order to explain the diversity and dynamics of life on the Earth. But Fox basically points out that there is another very powerful force at work: the flow of energy. Even without natural selection, the flow of energy alone would result in a diversity of forms of life.

A Hierarchy of Lives

The reason why so many theories tend to identify the second law of Thermodynamics as the principal driving force of biological order is that it is the only physical law that distinguishes between past and future, the only one that can explain irreversible processes, such as evolution and growth. The temptation is irresistible. But the true implications of the law of entropy (and even the very definition of entropy) are far from being well understood.

Modern physics adheres to the Cosmological Principle that the average properties of the universe are the same everywhere and in every direction, that the laws of nature are invariant under translations and rotations, i.e. the universe is "isotropic". The US physicist David Layzer ("Cosmogonic Processes", 1968) turned it into the "Strong Cosmological Principle": there was nothing in the initial state that gives a position or a direction a preferred status. It follows from this principle that randomness is a property of the universe. "A complete description of the universe contains only statistical information", and the probabilities that figure in that "statistical information" simply reflect the absence of microscopic order (not our ignorance of the truth), and this absence of microscopic order is responsible for the decline of macroscopic order embedded in the second law of Thermodynamics.

The "Strong Cosmological Principle" implies cosmological indeterminacy, i.e. that the more you know about the macrostate of the world the less you know about the microscopic states that cause it (a kind of indeterminacy related to Heisenberg's principle of indeterminacy). In other words, knowledge of the macrostate does not include knowledge of its corresponding microstates. Einstein's interpretation of the probabilistic nature of Quantum Mechanics was similar: Quantum Mechanics is a statistical theory useful to describe sets of particles, but not individual objects. Likewise, Layzer believes that we can only describe statistical properties. Layzer, however, parts from Einstein in explaining the origin of this limitation: Einstein thought that it was simply due to our ignorance of the fundamental laws of nature, whereas Layzer thinks that it follows from the equivalence of all positions and all directions anywhere in the universe.

In theory, as Clausius and Boltzmann realized, the universe should be doomed to decline towards states of decreasing order because natural processes generate more and more entropy. One would conclude that at the beginning the universe was more orderly than it is now. Layzer, however, points out that this is not necessarily true in an expanding universe: order is created by the cosmic expansion. If entropy in the environment increases more than the entropy of the system, then the system becomes more ordered in that environment. Entropy and order can both increase at the same time without violating the second law of thermodynamics.

Unlike Prigogine, Layzer does not need to assume that an energy flow from the environment of a system can cause a local decrease in entropy within the system. Entropy and order increase together because, technically speaking, the realization of structure lags behind the expansion of phase space.

Layzer does not believe that the universe was born complex and hot. He believes that such a vision leads to blatant contradictions, whereas an initially structureless and cold state accounts for the features of the universe that we observe today, starting with gravitational clustering, as that kind of state inevitably leads to a hierarchy of self-gravitating clusters.

The alternative to Boltzmann's model (that the universe began in a highly ordered state) is a model in which the initial state had no order at all and order (the hierarchy of self-gravitating clusters) was created as it cooled down and density fluctuations got amplified, each self-gravitating assembly becoming a component to generate with other peers the next self-gravitating assembly. Order was created because a gravitating gas with negative internal energy in an expanding medium is inherently unstable.

Layzer believes that order must be given the same status as energy among the fundamental features of the world, and in that case living organisms have something that nonliving organisms don't have: a special kind of organization. Living matter is made of the same stuff as nonliving one, but they way it is organized is different. In his opinion this approach can help solve the mystery of why evolution tends to create more complexity. In theory there should be no average improvement or decline in complexity. In practice there has been a vast improvement from the first living cell to today's life. He quotes Peter Medawar: "The inexplicable tendency of organisms to adopt ever more complicated solutions to the problem of remaining alive" ("A Biological Retrospect", 1982).

The fundamental property of living organisms is, to Layzer, "reproductive instability", a property that turns life into a kind of viral infection: "Just as a virus transforms its living host into a factory for producing more virus, so life transforms its inert host into a factory for producing more life". He traces this reproducing instability back to an inherent instability of genetic material, and this one back to a property of all molecules organized in a particular way.

Evolution is due to the cooperation of two processes: variation and selection. Genetic variation allows new forms of life to emerge and natural selections determines which ones survive and reproduce. In 1949 the Russian zoologist Ivan Schmalhausen proposed that evolution is a process of hierarchical construction leading to a functional hierarchy in which each functional unit is created by aggregation of preexisting functional units. Genetic variation is the consequence of this general process of hierarchical construction, and evolution shaped this hierarchical process. Therefore, Layzer argues that genetic variation, that indirectly shapes evolution, is not a random process but was shaped by evolution itself. Because it is not random, but driven by a hierarchical process, it ends up fostering complexity, and the apparent paradox is solved.

In detail, Layzer distinguishes alpha and beta genes. Alpha genes are the genes of classical genetics and the regulatory genes that transcribe these genes into RNA. Beta genes are not directly involved and serve maintenance chores. Layzer's theory is that beta genes drive evolution by promoting mutations that are likely to increase fitness and suppressing mutations that are likely to decrease fitness. He then argues that hierarchical construction is the natural consequence of this system.

Biological order is not only created by evolution, it is also created by development. Biological order due to evolution is embodied in DNA. Biological order due to development is embodied in neural connections. Layzer adopts the stance of ecological realism, that perception is an active process during which we construct our cognitive life. That too is a hierarchical structure.

Inside the brain Layzer finds again a hierarchy, this time a hierarchy of neural circuits modeled after Alexander Luria's three functional units (1942).

His "Strong Cosmological Principle" has made randomness a property of the universe: the universe began from a state of pure randomness (zero order). He uses that randomness to prove that creative processes such as biological evolution (as well as cultural evolution) inevitably generate order in an unpredictable way. The future (both of life and of human behavior) is not predetermined.

The Irreversibility of Life

Not everybody agrees with Prigogine's view of living systems as dissipative structures and with Schroedinger's view of life as "negentropic".

A law known as "Dollo's law" states the irreversibility of biological evolution: evolution never repeats itself. Darwin's natural selection does not necessarily prescribe progress or regression, does not imply a direction of evolution in time, it only states an environmental constraint. Indirectly, Dollo's law does: it prescribes a trend towards more and more complex, and more and more ordered, living structures. Dollo's law expresses the visible fact that reproduction, ontogeny and phylogeny are biological organizations whose behavior is irreversible: both during growth and during evolution. Entropy of biological information constantly increases. We evolved from bacteria to humans, we grew from children to adults.

The goal of the unified theory of evolution put forth in the 1980s by the Canadian biologist Daniel Brooks and the US ecologist Edward Wiley is to integrate this law with natural selection.

Unlike Prigogine, Wiley and Brooks believe that biological systems are inherently different from dissipative structures. Biological systems, unlike physical systems, owe their order and organization to their genetic information, which is peculiar in that it is encoded and hereditary. Dissipation in biological systems is not limited to energy but also involves information, because of the genetic code, which is transmitted to subsequent generations. Organisms simply live and die, they don't evolve. What evolves is the historic sequence of organisms, which depends on genetic code. The genetic code must therefore be placed at the center of any theory of evolution.

Unlike most theories of information, that use information to denote the degree to which external forces create structure within a system, Brooks-Wiley's information resides within the system and is material, it has a physical interpretation. It resides in molecular structure as potential for specifying both homeostatic and ontogenetic processes (processes for, respectively, maintaining internal equilibrium and growing). As the organism absorbs energy from the environment, this potential is actualized and is "converted" into structure.

What they set out to prove (following Lotka's original intuition) is that evolution is a particular case of the second law of Thermodynamics, that Dollo's law is the biological manifestation of that second law. Biological order is simply a direct consequence of that law. The creation of new species is made necessary by the second law and is a "sudden" phenomenon similar to phase changes in Physics. Phylogenetic branching (the creation of new species) is an inevitable increase in informational entropy.

In this scenario, the interaction between species and the environment is not as important in molding evolution: natural selection mainly acts as a pruning factor.

Over short time intervals, biological systems do behave like dissipative structures. But over longer time intervals, they behave like expanding phase space systems (as proved by Layzer). Their relevant phase space is genetic, an ever increasing genetic phase space.

The Brooks-Wiley theory is Darwinian in nature, as it subscribes to the basic tenet that evolution is due to variation and selection, but, in addition, it also allows the possibility for evolution to occur without any environmental pressure.

The Origin of Form

In the 1890s, the German physiologist August Weismann realized that living organisms exhibit a dichotomy: a piece of the organism's machine is designed to reproduce and another piece is designed to achieve form. One needs both in order to specify an organism.

In a sense, Weismann was the Descartes of biology: he separated reproduction and "soma" (form), genetics and morphogenesis. So one can study genetics without studying development, and viceversa. He understood that the problem has to be reframed in terms of "information".

One of the most puzzling features of life is, indeed, development. During development, cells split and split and split. Every time a cell splits, the new cells inherit (almost) exactly the same genes. But then, mysteriously, some cells become liver cells and some cells become bone cells and some cells become blood cells. Somehow a cell knows which proteins it has to make, and when and how many of it. If they run the same program, how come that two cells become two different things? And how do they know the position where those two things have to be?

Not surprisingly, Weismann concluded, logically, that each cell must include a different set of genes. We have, instead, learned that each cell includes the same set of genes, but a cell's genes are "regulated" in such a manner that only some are active. Each cell has the same set of genes, but each cell has different genes active. Nonetheless, the question remains: what determines which genes are switched on and off in a given cell? How do "regulatory" genes know that a cell has to become part of a hair rather than a liver? Whatever the mechanism, it must be extremely reliable, because billions of humans get eyes in their face and not in their feet. Even more striking is the fact that zebras have black and white stripes: how do those cells know that they have to be white or black? And what causes the skin of the zebra to have stripes, rather than a uniform color?

The British mathematician Alan Turing proposed a solution based on the properties of standing waves ("The Chemical Basis of Morphogenesis", 1952). His equations of "intercellular reaction-diffusion" show that patterns such as the spiral of a seashell, the stripes of a zebra or the spots of a leopard can originate from the local interactions of two chemicals ("morphogens") that spread into the body at different rates.

The British biologist D'Arcy Thompson argued that genetic information alone does not fully specify form. Form is due to the action of the environment (natural forces) and to mathematical laws. Form arises because of mathematical and physical properties of living matter, just like the shape of nonliving matter. D'Arcy had discovered an interesting property of life. Many animal forms can be shown to be homologous on a warped Cartesian diagram: bend the space a bit and the shape of an animal becomes the shape of another animal. Form is a mathematical problem, and growth is a physical problem. The form of an object is a resultant of forces. By simply observing the object, we can deduce the forces that have acted or are acting on it. This is easily proved of a gas or a liquid, whose shape is due to the forces that "contain it", but it is also true of solid objects like rocks and car bodies, whose shapes are due to forces that were applied to them.

The formative power of natural forces expresses itself in different ways depending on the "scale" of the organism. Mammals live in a world that is dominated by gravity. Bacteria live in a world where gravity is hardly visible but chemical and electrical properties are significant.

Ultimately, D'Arcy believed that living organisms owe their form to a combination of internal forces of molecular cohesion, electrical or chemical interaction with adjacent matter, and global forces like gravity.

Differentiation

Genes, obviously, do not carry all the information needed for an organism to develop.

How do cells of many different kinds come to occupy the "right" position in space? How do brain cells grow in the brain rather than, say, below the armpit? The phenomenon is even more mysterious because we now know that the early embryos of many animals, from insects to mammals, exhibit the same spatial pattern of activity of the same group of genes, before a morphological structure is created.

A body is shaped by the orderly movement of billions of cells to the locations that specify their role. Cells are not genetically programmed to perform a specific role, but during development they become specialist. It appears that what a cell will do for the rest of its life depends on where its journey ends. "Growth" is this mass migration of cells towards an unknown destination that will determine their future.

"Pattern formation" is the mechanism by which cells in different parts of a developing organism acquire different fates. "Pattern formation" constitutes the main concern of Developmental Biology. Today, we believe that an organism is made from a very large number of autonomous cells which can interact among each other and that the whole functional organism "emerges" (i.e., arises) from local interaction of cells.

Little is known about the physical process that allows this to happen, but cells in the embryo appear to be able to regulate their adhesion to surfaces and to other cells and they appear to do this to change shape or move.

Self-formation

Information-based hypotheses abound, and the very first one was advanced by Alan Turing in person: a uniform distribution of chemicals can develop spontaneously in a wave of regularity. This would explain, incidentally, why Nature prefers repeated patterns.

The US biologist Stuart Kauffman views the problem of cell differentiation as a problem of networks that search for stability. Each cell is equipped with the same network of genes, but the process that is occurring within each network is different: different genes are active in different cells. There is an almost infinite number of combinations in which genes can be active in a cell, but only a few of these combinations (precisely, the square root of the number of genes) correspond to mathematical "attractors". In the imaginary landscape of all possible genetic processes (the epigenetic landscape), there are basins of attraction. Those attractors correspond to the cell types that will arise.

The US biologist Brian Goodwin believes that one cannot make sense of nature simply based on the information carried by genes. Nature is more easily explained by the self-organizing patterns of networks. There are physical laws that originate a universal tendency towards complex adaptive systems. Genes carry instructions, but those instructions are subject to that intricate web of constraints that is the environment.

Topobiology

The US biochemist Gerald Edelman explains location-dependent development of body cells (e.g., how a cell knows where in the body it is supposed to grow in order to generate the shape and function of the animal) by assuming that development is based on topo-biological events which are regulated by cell-adhesion and substrate-adhesion molecules on the surface of the cell. In other words, a cell's competence is due essentially to its location.

In detail, the story reads like this. Living systems exhibit three properties that allow them to exist: heredity, variation in their hereditary material, and competition as the environment changes. Living systems are self-replicating systems, whose genome undergoes mutation and whose variant individuals undergo natural selection. Characteristic of living systems is development, in particular morphogenesis, the emergence of form during embryonic development. Roughly the same cell types appear in different parts of the body. The difference in position and shape results from the interaction of a number of driving forces (namely cell division, cell motion and cell death), which determine the number of cells in a particular region, and regulatory processes (namely cell adhesion and cell differentiation), which determine the interaction among cells.

Pattern, and not mere cell differentiation, is the evolutionary basis of morphogenesis.

The cell surface, not its core, plays the fundamental role in this process, because it mediates signals from other cells and links with other surfaces to form tissues. A sequence of interactions between certain special types of genes via epigenetic signal paths provides the basis of pattern by controlling temporal sequences of mitosis, movement, death and further signaling.

Epigenetic landscape

An apparent paradox is that different genetic programs can produce the same organism. In most cases, far less than 50% of the genes of an individual are shared with individuals of the same species. The individuals of a species differ in all sorts of ways, but somehow their genetic programs are tolerant to such differences and eventually yield individuals of the same species. In the 1950s the British genetist Conrad Waddington proposed a possible solution to the apparent paradox: the development of an individual is immune to the pull of the genes. Development is "canalized". He imagined an "epigenetic landscape" created by the concurrent pressures of the environment and the genetic program. Development occurs as a traversing of this landscape. The landscape varies from individual to individual, but it always maintains its fundamental shape of a gently sloping surface, that ends in the same valley. No matter how the landscape is traversed, the motion will always end in that valley.

Catastrophes

Rene Thom, the French mathematician who "invented" catastrophe theory, assumed that the fundamental problem of biology is a topological problem: how form is built.

The biochemistry of life should be explained by morphogenesis, not the other way around.

Catastrophe theory is basically a classification of the ways in which forms can change into other forms. Morphogenesis is due to the disappearance of some attractors (in the epigenetic landscape) and the capture by new attractors, i.e. the new form.

Death is easily defined: the transformation of a metabolic field into a static field. On the contrary, the birth of life would require an "infinite" number of local transformations in order to achieve the "anabolic" transformation from static to metabolic (from simple ingredients to the complex structure of living tissue).

Furthermore, once life occurs it is not clear why it stops at all: the underlying processes are reversible, therefore life should continue forever.

Vitalism

Genes do not carry all the information needed to specify the development of an organism. The same genetic program in two cells yields a blood cell and a liver cell. Somehow there must be other "information" available that tells one cell to become a blood cell and the other to become a liver cell. One clue to the solution of this mystery is that, as cells differentiate within the organism, different genes are "expressed" in different cells.

At the end of the 19th century, the German embryologist Hans Driesch realized that a mutilated embryo would still develop into a fully-functioning living organism. He could not find any rational explanation and posited the existence of a "life force", or "entelechy". This was a variation on the old theory of "vitalism": that organic matter is fundamentally different from inorganic matter due to the presence of a vital principle.

Driesch's entelechy was a goal-directed (or "teleological") organizing process that would guide morphogenesis, regardless of any other information. Entelechies are organized in hierarchies (so that one doesn't need an entelechy for every single organism that can possibly exist).

Then in the 1930s biologists such as Paul Weiss and Hans Speman (the first one to envision cloning by transferring the nucleus from one cell to another) hypothesized that "morphogenic organizing fields" helped organisms take their shape. The British geneticist Conrad Waddington gave these fields a mathematical meaning with "chreodes", developmental pathways (channels) in his epigenetic landscape: form follows the channels rather than wander in other parts of the landscape.

The German physicist Walter Elsasser concluded that Physics is not enough to explain life, and proposed the expansion of Physics to "biotonic" laws.

The US mathematician Ralph Abraham ("Vibrations and the Realization of Form", 1976) introduced a similar notion, that of "macrons": a macron is a collective vibrational pattern (many things that start vibrating together in synchrony). Abraham showed that macrons are ubiquitous in nature (in solids, liquids and gases).

The British physicist Paul Davies also resorted to a sort of life force in order to explain the origin of life, but this "life force" is, in his opinion, a kind of software program. Davies thinks that science must accept "information" as a fundamental quantity of the universe, that can be traded by "informational" forces the same way that matter is traded by physical forces. The natural laws of informational forces must be compatible but not reducible to the laws of physical forces.

The Memory of Nature and Morphic Fields
The British philosopher and biochemist Rupert Sheldrake offered a neo-Aristotelian view of life and nature.

Sheldrake views the growth of form as one of the fundamental processes of Nature. The foundation of Sheldrake's concept of "formative causation" is the idea that memory is inherent in Nature (an idea borrowed from the nineteenth century biologist Samuel Butler).

Natural systems inherit a collective and cumulative memory from all previous systems of their kind, regardless of time and space separation; and natural systems in turn contribute to the growth of this collective and cumulative memory. Habits are inherent in the nature of all organisms because of the memory that organisms inherit from previous organisms of the same kind. For living organisms, not only genes are inherited, but also habits, which include development habits such as morphogenesis (the growth of form).

The universal memory expresses itself through "morphic fields". Morphic fields are an organizing principle of Nature. A morphic field is a field (or pattern or order or structure) of form. Such fields have a kind of built-in memory derived from previous forms of a similar kind.

Each natural system has its own morphic field that shapes its behavior. There is a morphic field for pears, whales, crystals, etc. There is a nested hierarchy of fields within fields. Morphic fields evolve by natural selection.

Morphic fields are responsible for form and organization (in biological as well as material systems). The morphic field of a system derives from morphogenetic fields associated with all previous similar systems (across both space and time).

"Morphic resonance" is the process by which the past becomes present, i.e. it is the process of transmitting formative-causal information across space and time. "Morphic resonance" is the process by which the form of a system is influenced by the forms of past similar systems through the morphic field. The morphic field influences the form of a system, and, in turn, the form of the system influences the field and thus any future form of similar systems. The more similar an organism is to previous organisms, the stronger it "resonates" with (learns from) them. Individual memory is simply self-resonance: an organism resonates with its own past, i.e. it "remembers" it. Conversely, there is a universal memory that all forms share, the form of all forms (which he compares to Bohm's "implicate order").

The persistence of the material form of a system depends on the continuous application of the morphic field on that system, which is, in turn, continuously recreated by morphic resonance.

By "form" Sheldrake means more than just shape: spatial order, including the internal structure. He points out that form is not matter/energy: the total amount of matter/energy in the universe is the same before, during and after the existence of an organism, but the existence of the organism causes a change in the way matter/energy is organized.

Form and energy are inversely proportional: energy expresses a principle of change, whereas form expresses resistance to change.

Formative causation implies, for example, that a new pattern of behavior should be transmitted across space and time to individuals of the same species: as individuals of the species learn something somewhere, other individuals of the same species, no matter where they are located, should be learning it too (to some extent).

Sheldrake believes that each species has its own fields, and within each organism there are fields within fields.

Within each of us there is the field for the brain and the heart; within are fields for different tissues inside these organs, and then fields for the cells, and fields for sub-cellular structures, and so on.

Such fields organize not only the fields of living organisms, but also the forms of crystals and of molecules. Each kind of molecule has its own kind of morphic field. So does each kind of crystal, each kind of organism, and each kind of instinct or pattern of behavior. These fields are the organizing fields of nature. There are many kinds of them, because there are many kinds of things and patterns in nature.

The development of organisms is regulated by such morphic fields, and so is the organization of behavior. Genes carry only a minuscule part of the biological information in nature. Most of inheritance depends on the memory which is carried within the organizing fields of an organism. This memory is a kind of cumulative memory which is constructed through a pool of species experience, depending on morphic resonance.

Genes do not carry all the information needed to shape an organism. Genes interact with the morphic fields of previous organisms of the same species.

Biological inheritance is about both genes and fields. Fields allow for Lamarckian inheritance of acquired characteristics. Inheritance of acquired characteristics occurs not because of transmission of genes but because of the effects of morphic fields, which are modified by individuals "learning" something and then influence the development of future individuals of the same species.

Memory is not stored in the brain but it resonates with the organism's own past. And a collective memory underlies our mental life (similar to Jung's collective unconscious).

Myths, rituals, traditions are expressions of that collective memory: morphic fields organize social and cultural patterns and through morphic resonance, rituals bring the past into the present, connect past individuals with present individuals.

Memories are never completely private. In principle, anybody could tune into our "private" memories and "read" our mind.

Sheldrake views all of Nature as a living organism. Nature is essentially "habit-forming", and all aspects of Nature are regulated by the principle of habit. The "laws of nature" are therefore better described as "habits of nature". The habits of animals and plants give them their habits of growth and their habits of behavior, or instincts.

(Sheldrake's theory, alas, flies against the evidence: children still need to re-learn how to walk and speak, despite the thousands of generations and billions of humans who did that in the past, a fact that sounds like overwhelming evidence that there is no morphic field for common behavior).

Beyond Chemistry: Tensegrity

In 1993 the US physician Donald Ingber popularized the concept of "tensegrity". Living systems, at all hierarchical levels, stabilize through the interplay of two forces, one which is tensional and one which is compressive. Ingber reasoned that, since cells continuously die, their chemistry alone cannot be responsible for the evolution for form. What is maintained is the architecture. Therefore, Ingber focused more on Architecture than on Biology. He re-discovered two types of structures that exhibit spontaneous and resilient stability: the geodesic dome invented by the US physicist Buckminster Fuller (in which the geometry of the components constrains the Physics of the components, thereby immobilizing the whole structure) and the "pre-stressed" sculptures built by the US artist Kenneth Snelson (in which rigid components tense flexible components and flexible components compress rigid components, thereby "pre-stressing" the whole structure). These "tensegrity" structures share the property of optimizing structural stability while minimizing building material.

Ingber proved that living cells (and, in particular, their internal framework, the "cytoskeleton") behave like tensegrity structures, and that principles of tensegrity also govern (at least) tissue formation. Geodesic forms abound in nature, from the cytoskeleton to some carbon atoms.

Ingber believes that tensegrity accounts for the continuity of movement: when an organ moves, millions of cells are affected, and each one has to adapt to the movement of the others. A tensegrity structure allows for a balanced transmission of tension to the elements of the structure and guarantees the "harmony" of the whole structure. In other words, the structure does not break or fall apart, but re-distributes tension and therefore redesigns itself.

Ingber believes that life began in layers of clay, a substance whose atoms are arranged geodesically and whose porosity allows for the catalysis of chemical reactions such as the ones that led to the building blocks of life. Life developed before any genetic mechanism was present.

Then DNA created a way to accelerate evolution. It is not a coincidence that pre-stressed and geodesic forms predominate in the living world.

The Function of Growth

The construction of form is, of course, only an aspect of growth. The basic question is: why do beings grow? What is the goal of growth? Why aren't we born as adults? Wouldn't it be simpler if we were born like adults and had to worry only about reproducing? We have to protect and nurture our offspring, which results in a great waste of energies and in lower survival rates. A species that did not need to grow would be highly efficient. Why do living things grow instead of being built?

There might be a few reasons. The first one has to do with complexity. It would require a huge amount of specification to assemble a body, cell by cell, whereas "growth" is a process whereby each component of the system helps specify the system as a whole. One needs very little to start, and there is virtually no limit to how far one can go.

A second advantage is that a system that is built from scratch is not as resilient, as easy to repair, as a system that has developed through a number of stages. Because growth is an on-going never-ending process, most faults (such as wounds, cuts, fractures) get repaired naturally. The system is tolerant to most faults and will still operate. In an artifact, most faults disable the entire system (a mere flat tire is enough to stop a very expensive vehicle) and may even destroy it (a mere washer was enough to blow up a space shuttle).

Finally, a growing being better integrates with the environment. Because growth depends on the surrounding matter as well as the genetic program (i.e., we eat plants and animals, we breathe air, etc.) the resulting body is better equipped to cope with the challenges posed by our environment. In a sense, there are no "unpredictable" events, all possible accidents have been implicitly predicted in the way our body grows.

Then the complementary question: why does growth stop? What is so special about our adulthood that makes it the terminal point of growth, after which decay begins? Why does growth end and fade into decay? When does decay really begin and why at that point rather than at any other point? And when does it really end? We know when a body is created, because all of a sudden we can see it and touch it, but we don't really know when a body is destroyed, because it fades away slowly. Needless to say, things would be even easier if organisms did not decay, if we just lived forever...

Interactivity

The view of the gene as a "ghost in the biological machine", as the set of instructions for building living beings, was criticized by the US philosopher Susan Oyama on the grounds that it perpetuates the misleading model of nature-nurture dualism (inherited versus acquired characters).

The western tradition assumes that form preexisted its appearance in bodies and minds (e.g., as a genome). Information is the modern source of form: ubiquitous in the environment as well as in the genome. The development of an organism is traditionally explained as a dual, parallel process: on one hand, translating information in the genome ("nature"); on the other hand, acquiring information from the environment ("nurture"). Both processes are dependent on information. Information therefore regulates development. This view has deep cultural roots, but Oyama objects that it is nothing more than myth. Oyama's viewpoint is that information (e.g., from the genome) is itself generated, it develops. Information itself undergoes a developmental process.

Opposed to both nurture and nature, Oyama argues that the form of an organism cannot be transmitted in genes or contained in the environment, and cannot be partitioned by degrees of coding: it is constructed during the developmental processes. Information in the genes and information in the environment are not biologically relevant until they participate in the processes that actually build form. Form emerges through a history of interactions at many hierarchical levels, and genetic form is but one of the "interactants". Form is the result of interactive construction, not the outcome of a preexisting plan. The distinction between inherited and acquired characters should be replaced by the notion of development systems.

An organism inherits its environment, as much as it inherits its genotype. It inherits some competence, but also the stimuli that make that competence significant.

Life and the Universe

Life appears less and less like a weird exception to the rules of Physics and more and more like a natural consequence of the way our universe works.

The British physicist Freeman Dyson was instrumental in building the field at the borders of physical, biological and information sciences. Inspired by the British physicist Jamal Islam, who calculated how matter would evolve in universes which expand forever, Dyson computed mathematically what life is and how it will evolve. A closed universe is doomed to collapse and life with it. Since a system's entropy is a measure of the number of alternative states of the system, the complexity of a living organism should be proportional to the negative of its entropy. Dyson even computed the entropy of a human being (the rate at which humans dissipate energy times the human body's temperature times the duration of a unit of consciousness): 10 to the 23th. Life is a form of order, and low temperature favors order. Life and intelligence are immortal, because sources of memory will grow constantly as the universe cools down. Interestingly, "life" for Dyson is not necessarily the stuff made of proteins. "Life resides in organization, not in substance".

As the US physicist Steven Frautschi, among others, noted, there is a striking parallelism between the evolution of the expanding universe and the evolution of life on Earth: because life on Earth has a steady free energy source (the sun), it does not need to come to equilibrium and may even evolve away from it (as it did when it created more and more complex beings, such as ourselves); because the universe has a steady free energy source (the uniform expansion itself), it does not need to come to equilibrium and may even evolve away from it (as it did when it created more and more complex clumps of matter, such as galaxies). Both biological evolution and universe evolution could turn out to be consequences of non-equilibrium processes.

The Omega Point

The "omega point theory" of the universe imagined by the US physicist Frank Tipler was meant as a rigorous mathematical proof of the existence of an omnipresent, omniscient and omnipotent god. Tipler even calculated the likelihood that every human being be eventually resurrected, and conjured up a physical model of life in heaven, hell and purgatory, all based on Information Theory, Quantum Mechanics and Relativity Theory.

His basic point is that life is "information coding" preserved by natural selection: a being is alive if it encodes information and such information is preserved over time by natural selection. Given this definition of life, it is possible to compute how much energy is sufficient and necessary to extend this process till the very end of time in a closed universe.

Furthermore, a simulated universe is, for all purposes, a universe, and its inhabitants are, for all purposes, as real as us, because there is no way that the simulated beings can realize they are simply being simulated. We may well be just that: simulations inside a computer. Virtual reality is no less real than actual reality, because there is no way to distinguish one from the other. In a sense, only virtual reality exists.

Tipler calculates the maximum amount of information needed to simulate brains and entire humans, by basing his conjectures on the Bekenstein bounds (the upper limit on information density, according to Quantum Theory).

The "omega point" is the final singularity of the history of a closed universe, the point of infinite information, which is neither space nor time nor matter, but is beyond all of these and experiences the whole of universal history all at once.

Tipler believes that our universe is a Taub universe, a universe which, after the expansion phase is over, will start contracting at different rates in different directions, thus leading to an oblate spheroid shape. In such a contracting universe, temperature will be higher in the "contracted" direction and the difference of temperature between that direction and the others will generate energy. The fact that a closed universe must be in an infinite singularity actually means that infinite energy will be produced.

In a Taub universe the differential collapse becomes a source of free energy, which life can use to survive forever. This source of energy will be to eternal life what the Sun is to life on Earth. The finite singularity of the universe and eternal life happen to coincide...

The Meaning of Life

A more scientific way of asking "what is the meaning of life" is: "what is responsible for my existence?" Which law in the universe has caused some molecules to assemble and become my body and then grow to the stage I am at now?

The law of entropy has widely been considered the "smoking gun" of the situation. Unfortunately, nobody seems to really know what "entropy" means. Macroscopically, entropy is the ratio of heat to temperature, which is not a very intuitive definition. Microscopically, it is the number of micro-states that implement a macro-state (Boltzmann's definition), an even less intuitive concept. The law of entropy is even less well understood. Macroscopically, it states that the entropy of the universe can never decrease. This statement is not very easy to relate to our daily lives. Its more microscopic formulation as "heat can never be completely converted into work" is much more useful for practical purposes. In other fields, it can be better understood as "order cannot be created, unless at the expense of creating disorder somewhere else" (as the chemist John Holmsted put it, "the creation of local order requires generation of global disorder").

Prigogine showed that the law of entropy is useful to classify in which ways a system can evolve: a "closed" system (one that is fully isolated from the rest of the universe) can only evolve towards increased entropy, i.e. increased disorder (but, needless to say, no system in nature is really closed); an "open" system that expels energy (or matter) can evolve to ever-higher levels of order in a state of equilibrium (a star is an example of such an open system); an "open" system that both expels and absorbs energy can evolve to ever-higher levels of order while always being far from equilibrium.

From this, one can perceive a similarity between the last category and living systems, and therefore be tempted to infer that living systems "are" in fact that category. One problem is that the complexity of living systems is not easily reduced to an abstract category. Another problem is that many physical systems belong to the same category that we would not like to consider living. A dishwasher absorbs energy/matter from an outlet and a pipe and expels energy in the form of hot water down a drain, but that doesn't automatically entitle it to the rank of living system.

But then a dishwasher was manually built, whereas living systems build themselves from virtually nothing. That "encoding" of information is really the clue to the meaning of life. That's why information has become more and more the focus of attention. What sets living systems apart from physical systems is not the flow of energy/matter: it is the fact that whatever they do is to some extent due to a "program". Living systems are machines programmed to perform the tasks of growth, reproduction and evolution. The downside of this argument is that any information-based simulation of life (including one performed into a computer) qualifies as life itself.

But this still doesn't answer the question: "which natural law is responsible for my existence?"

Any living system is built on top of physical systems, of matter, of "stuff". There is no reason why its natural laws should be any different than the natural laws that work on stuff. We feel that the answer to that question must lie in a general property of the universe. Possibly the reason Physics still doesn't know the answer is that Physics still doesn't know the general properties of the universe.

Universal Darwinism

In view of recent progress in several disciplines, in which the Darwinian paradigm keeps recurring at different levels of organization (species, immune system, brain), it seems reasonable to assume that Darwinism is a universal principle, not limited to Biology. The British psychologist Henry Plotkin actually believes that Darwinism is likely to become the basis of all science, the idea spreading beyond biological evolution.

"Universal Darwinism" will be a theory based on Darwinism but general enough to encompass everything. A likely candidate structure for an empirical science is one based on the concepts of replicator (an entity that can make copies of itself) and interactor (an entity that can propagate replicators in space and conserve them in time while interacting with the environment). The presence of this combination is evidence that evolutionary algorithms are at work. They occur in life, in the brain, in the immune system, in memes.

Further Reading

Abraham, Ralph: ON MORPHODYNAMICS (Aerial Press, 1985)
Atkins, Peter-Williams: THE SECOND LAW (WH Freeman, 1984)
Bateson, Gregory: MIND AND NATURE (Dutton, 1979)

Brooks, Daniel & Wiley E.O.: EVOLUTION AS ENTROPY (Univ of Chicago Press, 1986)

Carvalo, Marc: NATURE, COGNITION AND SYSTEM (Kluwer Academic, 1988)

Davies, Paul: THE FIFTH MIRACLE (Simon & Schuster, 1998)

Driesch, Hans: SCIENCE AND PHILOSOPHY OF THE ORGANISM (1908)

Dyson, Freeman: INFINITE IN ALL DIRECTIONS (Harper & Row, 1988)

Elsasser, Walter: THE CHIEF ABSTRACTIONS OF BIOLOGY (1975)

Fox Ronald: ENERGY AND THE EVOLUTION OF LIFE (Freeman, 1988)

Ganti ,Tibor: THE PRINCIPLE OF LIFE (Omikk, 1971)

Goodwin, Brian: HOW THE LEOPARD CHANGED ITS SPOTS (Charles Scribner, 1994)

Hutchinson, Evelyn: THE ECOLOGICAL THEATER AND THE EVOLUTIONARY PLAY (1965)

Islam, Jamal: THE ULTIMATE FATE OF THE UNIVERSE (Cambridge Univ. Press, 1983)

Jansch, Erich: THE SELF-ORGANIZING UNIVERSE (Pergamon, 1980)

Kauffman, Stuart: THE ORIGINS OF ORDER (Oxford University Press, 1993)

Kestin, Joseph: A COURSE IN THERMODYNAMICS (Blaisdell, 1966)

Kuppers, Bernd-Olaf: INFORMATION AND THE ORIGIN OF LIFE (MIT Press, 1990)

Langton, Christopher: ARTIFICIAL LIFE (Addison-Wesley, 1989)

Layzer, David: COSMOGENESIS (Oxford University Press, 1990)

Lotka, Alfred: ELEMENTS OF MATHEMATICAL BIOLOGY (Dover, 1925)

Margalef, Ramon: PERSPECTIVES IN ECOLOGICAL THEORY (Univ of Chicago Press, 1968)

Maynard-Smith, John: EVOLUTIONARY GENETICS (Oxford University Press, 1989)

Mikulecky, Donald: APPLICATIONS OF NETWORK THERMODYNAMICS PROBLEMS IN BIOMEDICAL ENGINEERING (New York University Press, 1993)

Morowitz, Harold: ENERGY FLOW IN BIOLOGY (Academic Press, 1968)

Morowitz, Harold: FOUNDATIONS OF BIOENERGETICS (Academic Press, 1978)

Morowitz, Harold: ENTROPY AND THE MAGIC FLUTE (Oxford University Press, 1993)

Murray, James Dickson: MATHEMATICAL BIOLOGY (Springer-Verlag, 1993)

Nowak, Martin: EVOLUTIONARY DYNAMICS (2006)

Odum, Eugene: FUNDAMENTALS OF ECOLOGY (1953)

Oyama, Susan: ONTOGENY OF INFORMATION (Cambridge University Press, 1985)

Plotkin, Henry: DARWIN MACHINES AND THE NATURE OF KNOWLEDGE (Harvard University Press, 1994)

Plotkin, Henry: EVOLUTION IN MIND (Allen Lane, 1997)

Prigogine, Ilya: FROM BEING TO BECOMING (W.H. Freeman, 1980)

Schneider, Eric & Sagan, Dorion: INTO THE COOL (Univ of Chicago Press, 2005)

Schroedinger Erwin: WHAT IS LIFE (Cambridge Univ Press, 1944)

Sheldrake, Rupert: A NEW SCIENCE OF LIFE (J.P. Tarcher, 1981)

Sheldrake, Rupert: THE PRESENCE OF THE PAST (Times Books, 1988)

Speman, Hans: EMBRYONIC DEVELOPMENT AND INDUCTION (Yale Univ Press, 1938)

Thom, Rene: STRUCTURAL STABILITY AND MORPHOGENESIS (1968)

Thompson, D'Arcy: ON GROWTH AND FORM (Cambridge University Press, 1917)

Tipler, Frank: THE PHYSICS OF IMMORTALITY (Doubleday, 1995)

Ulanowicz Robert: GROWTH AND DEVELOPMENT (Springer-Verlag, 1986)

Weber, Bruce, Depew David & Smith James: ENTROPY, INFORMATION AND EVOLUTION (MIT Press, 1988)

Weismann, August: THE GERM-PLASM (Scribner's, 1893)

Weiss Paul: PRINCIPLES OF DEVELOPMENT (Holt, 1939)

Weizsacker, Carl-Friedrich von: DIE EINHEIT DER NATUR (1971)

Woltereck, Richard: GRUNDZÜGE EINER ALLGEMEINEN BIOLOGIE (1932)

Wicken, Jeffrey: EVOLUTION, INFORMATION AND THERMODYNAMICS (Oxford Univ Press, 1987)

Woese, Carl: THE GENETIC CODE (Harper & Row, 1967)

ALTRUISM: FROM ENDOSYMBIOSIS TO SOCIOBIOLOGY

Competition vs Cooperation

Darwin popularized the paradigm that competition among living beings ("survival of the fittest") is the force that shapes evolution.

This is not what we commonly observe. Nature is certainly cruel with the weak, but the weak have a tendency to pool together and fight adversity. The whole concept of "society" stems from the biological instinct to team together.

Why would an individual help another individual of the same race, thereby violating the principle of "survival of the fittest" and lowering its own chances of survival?

Kin Selectionism

One possible answer is "kin selectionism", according to which selection operates at the level of kin, of closely related individuals, and not only at the level of the individual.

The British biologist John Haldane ("Population Genetics," 1955) pointed out that altruism is proportional to genetic proximity. I share genes with my brother, and therefore I am willing to help him survive. I share genes with cousins too, and therefore I am still willing to help them, but I share less genes with them than with my brother so I am less motivated to help them rather than my brother. It is not the survival of the individual that matters: it is survival of as many genes as possible. Parental care for offspring has, therefore, a genetic explanation.

The British biologist William Hamilton ("The Genetical Evolution of Social Behaviour I and II", 1964) argued that altruism too evolved by natural selection for a utilitarian reason: altruism helps genes as a global pool, even if at the expense of the survival of a specific individual. Altruism is just another step, beyond personal survival and reproduction, in the program to maximally proliferate the genes of an organism.

Traditionally, selection (and therefore evolution) had been viewed as driven by "reproductive success"; but Hamilton, armed with mathematical tools, extended that concept to the reproductive success of close relatives (or "kin"). Every individual has an investment in its own genetic pool. The investment peaks in its own body, but it is not limited to the body, it extends, albeit in lesser amounts, to all of its kin, and it is proportional to how closely they share the same genes. Hamilton captured the essence of kin selection (later better known as "inclusive-fitness theory") in a simple equation that defined mathematically the concept of "inclusive fitness". The equation shows that it benefits an individual to aid kin in order to promote its "inclusive fitness". In other words, the individual is programmed to preserve not only itself but also other individuals that share a similar genetic repertory, in a manner proportional to that similarity.

Hamilton's theory provided an explanation for why parents care for their offspring and why females are "choosier" than males about their mates. Cooperation turns out to be but another aspect of competition.

Hamilton argued that selection operates at the level of the genetic pool, not at the level of the single genome.

In a sense, a family is but one organism with many organs, each member of the family being an organ. Each organ works with the others to keep the organism (the family) alive. If an organ dies but helps the others survive, it helps the organism survive. Altruism is not in contrast with Darwinism.

The US biologist George Williams then explained how evolution extended altruism beyond kinship: an individual's chances of survival are increased by having friends and decreased by having enemies. That simple. Thus it makes sense for any individual to maximize friendship and minimize antagonism. There is no need for conscious calculation: evolution has endowed individuals with "altruistic" instincts and emotions because it helps them survive. Most of the individuals who didn't have them did not survive to make children. Williams thus explained the Darwinian value of friendship.

Selfish Altruism

The US biologist Robert Trivers noted that there was more than cooperation at work. According to William Hamilton's genetic metrics, a child should see herself twice more valuable than her siblings. The parents, on the other hand, should see all siblings as equally valuable. Thus it is not surprising that siblings compete and fight for parental resources, while parents teach them to share equally. Parents have to literally brainwash their children into thinking that it is in their (each child's) interest to care for their siblings when in fact their genes tell them (the children) the exact opposite.

Beyond family, there is in general a whole repertory of attitudes that serves the purpose of regulating altruism (gratitude, compassion, trust, guilt, even hypocrisy). Eventually, it all boils down to game theory: how to maximize the chances of success and minimize the chances of failing.

We seem to be even equipped with a repertory of skills to lie, cheat and deceive, and we use that repertory to complement the equation that maximizes our chances of success, depending on social conditions. Our conscience is malleable, which is another way to say that our altruistic strategies are flexible. In a sense the reason why children lie is that they are just practicing the art of cheating. In fact the tendency in children to lie is so strong that they will stop lying only if punished consistently and severely. Otherwise the tendency to lie will amplify. Conscience is an adaptation of one's altruistic and anti-altruistic instincts to a specific social environment.

Group Selectionism
William Hamilton's theory of kin selection explained only why animals assist close relatives (by placing the emphasis on the genes that are shared by relatives). But not why we would help friends or even total strangers.

At the beginning of the 20th century the Russian philosopher Petr Kropotkin first campaigned the notion that animals must be social and moral. His view was not one of individual struggle for survival, but one of the struggle for survival by masses of individuals, a struggle not against each other but a collective struggle against the common enemy, i.e. the adversities of their environment. Cooperation is more important than competition.

Meanwhile, the Japanese primatologist Imanishi Kinji was arguing that cooperation is more important than competition in nature. Individuals form societies and cannot exist outside societies because it is through societies that they can solve the needs required to their survival.

The British zoologists Vero-Copner Wynne-Edwards argued in favor of group selection because he found evidence that it is groups (rather than single individuals) that adapt to the environment.

The US biologist David Sloan Wilson ("A theory of group selection", 1975) resumed that explanation of altruism and made a case for the evolution of altruistic behavior. His studies gave credibility to the theory of "group" selection. A group is not necessarily a group of kin, but can be any community of genetically unrelated individuals and even of different species (as in the case of symbiosis). A group is just analogous to an organism. After all, an organism can be viewed as a collection of genes that work together towards maximizing their common chances of survival. The same principle applies to a group, where individual genes are replaced by organisms, by collections of genes. Groups often behave like organisms. Such is the case with beehives, ant colonies, flocks of birds, schools of fish, herds and even human clans.

Selection may operate at many different levels, but certainly for some species, especially humans, living in a group, and helping each other, has provided a tremendous evolutionary advantage. While the idea of a "group" of altruistic individuals, who accept to live in hives, herds, clans at the expense of their own fitness, may sound antithetical to Darwin's principle of competition, it does make sense, precisely from the point of view of "fitness". Being part of a group may increase the chances of being "fitter" and therefore survive.

Robert Trivers' theory of "reciprocal altruism" ("The evolution of reciprocal altruism", 1971) explained altruism as founded on the idea of exchange: i help you and you will help me. He proved that individuals can benefit in the long term by trusting each other. In other words, altruism is actually selfish. Building on Trivers' theory, the Dutch zoologist Frans de Waal argued that communities yield benefits to the individual, and that is the biological reason the individual will try to promote the community. Human morality is based on the idea of exchange. A society always relies, to some extent, on altruism: a member must be willing to sacrifice part of her individuality in order to be part of a society, which, in turn, increases her chances of survival.

Games
Game theory, introduced by John Maynard-Smith ("The Logic of Animal Conflict", 1973), helps to explain how altruism evolved. Over the long term, non-zero sum games ("cooperative" games in which both players stand to win or lose) tend to have more positive outcomes than negative ones. In particular, one can devise strategies that will greatly enhance the players' outlook in the long term. Thus it is not surprising that everything from ecosystems to human societies are built on altruism. (By contrast, "competitive" or "zero-sum" games represent a relatively static world).

The most famous of non-zero sum games is the "prisoner's dilemma", in which two prisoners are offered (independently) the same deal by the prosecutor. If one confesses and the other does not, the former goes free and the other one gets the maximum sentence. If they both confess, they both get a medium-length sentence. If neither confesses, they both get a minor sentence. This is a game that can be played only once. But imagine a similar game that could be played thousands of times with thousands of players, each player using a different strategy. Game theory proves that there is indeed a best strategy to play this game.

John Maynard-Smith's use of game theory decoupled kinship and cooperation: individuals cooperate not because they share genes but because cooperation is the best strategy (and it has little to do with moral "altruism").

The US political scientist Robert Axelrod held a tournament of computers programmed to play the game each against everybody else ("The Evolution of Cooperation", 1981). The "winner" (the one that did best over the long run), equipped with the program "Tit for Tat" written by Anatol Rapaport, was also the simplest one: it cooperated with the computers that had cooperated in the past, and cheated computers that had not cooperated in the past (basically, it did to others what others had done to it). "Tit for Tat" was creating an ever more cooperative society. It used the simplest algorithm, and it yielded the best outcome. Nature likes that combination. Even if individuals do not communicate, they will tend to cooperate, simply because, over the long term, it is the best strategy.

The Austrian mathematician Karl Sigmund and the Austrian biologist Martin Nowak ("Evolutionary Dynamics of Biological Game", 2004) came up with mathematical descriptions ("evolutionary dynamic models") for five mechanisms for the evolution of cooperation: kin selection, group selection, graph selection, direct reciprocity and indirect reciprocity. These models show that competition leads to cooperation. Nowak's theory, in particular, is that the Prisoner's Dilemma, when played over and over, generates cycles from selfishness to increased altruism and back to selfishness. Nowak argues that most of the great innovations of life, and notably human language and cognition, are due as much to cooperation as they are to Darwin's variation and selection.

The theory of kin selection is weak because the evidence does not support it: eusocial species are rare (basically humans, ants and a few others) while kin selection predicts that most species should evolve social skills (especially in species for which genetic similarity of kin is very high). The Romanian mathematician Corina Tarnita showed ("The evolution of eusociality", 2010) that the very mathematics behind kin selection could be wrong. Building on her findings, Edward Wilson proposed that altruism is due to social genes. Within any given group the selfish are more likely to succeed, but groups of altruists have an advantage over groups of selfish people. This led to the evolution of eusocial species that are genetically programmed to cooperate. Group selection leads to "virtue", individual selection leads to "sin".

The Neural Correlate of Altruism
It is debatable whether there is a neural predisposition for being nice to others, i.e. whether there is something about the human brain that makes children altruists instead of selfish. Children are, of course, influenced by the teachings of their parents, and eventually learn that there is a reward for being nice (first of all to their parents and siblings, then to their neighbords and so forth). However, there is evidence to the contrary: siblings who presumably have similar brains can turn out to be wildly different in the way they behave towards others (one can be extremely selfish in a family of very generous people or viceversa).

The Origin of Sex
The classical explanation for the existence (and widespread existence) of sex in nature was given by the German physiologist August Weismann in 1889 ("The Significance Of Sexual Reproduction In The Theory Of Natural Selection"): sex increases variation which is then used by natural selection to improve the fitness of the species. Basically, sex accounts for faster rates of adaptation.

However, there is a component of altruism in this purely statistical game.

The US zoologist Alison Jolly contends that altruism is a fundamental aspect of evolution. The very existence of sex as a means of reproduction is proof that cooperation is a crucial evolutionary force. Sex is a trade-off: a genome sacrifices a part of its genes to team up with another genome and increase its chances of survival in the environment.

The British biologist Matt Ridley thinks that evolution is accelerated even by apparent enemies like parasites. Organisms adopted sexual reproduction in order to cope with invasions of parasites: parasites have a harder time adapting to the diversity generated by sexual reproduction, whereas they would have devastating effects if all individuals of a species were identical (if the children were as vulnerable to the same diseases as the parents). Co-evolving parasites help improve evolution because they force individuals to cooperate. The lesson to be drawn is that (the need to fight) competition often leads to cooperation. On a large scale, life is a symbiotic process that is triggered by competitors. And, of course, plants reproduce with the help of insects. Well over 300,000 species of plants may have been created by co-evolution with their pollinators. Cooperation is pervasive, both within a species and across species.

The emphasis in evolutionary theories has traditionally been on competition, not cooperation, although it is through cooperation, not competition, that considerable jumps in behavior can be attained.

In a sense, humans have mastered altruism the same way they mastered tools that allowed them to extend their cognitive abilities. Humans are able to deal with large groups of non-relatives. De facto, those individuals are "used" as a tool to augment the mind: instead of having to solve problems alone, the mind can use an entire group.

Endosymbiosis

The mechanism proposed by Darwin to explain the evolution of life on Earth is based on a delicate balance between a positive process, that of variation, and a negative process, that of selection. The inconsistencies encountered so far in the fossil record all seem to point towards a need for a stronger positive process, one that allows for a species to be born in far shorter times than the evolutionary times implied by Darwin's theory. It is true, as Michael Behe noted, that an organism is way too complex to be built by refinements, and it is true, as Stephen Jay Gould claimed, that species appear all of a sudden. Selection does account for the disappearance of variations that are not fit, but variation alone (and the set of genetic "algorithms" that would represent it) is hardly capable of accounting for the extraordinary assembly of a new organism. A more powerful force must be at work.

When we find that force, we may finally write the last chapter of "The Origin of Species", which Darwin never even tried to write: we still don't know how species originate.

That force may be hidden in the process of endosymbiosis, the process by which a new organism originates from the fusion of two existing organisms, or, more precisely, by which two independently evolved organisms become a tightly coupled system and eventually just one organism. "Endosymbiosis" is the process by which a being lives inside another being.

"Structural coupling" of organisms has been shown to be an accelerating factor in evolution both by the Chilean neurobiologist Humberto Maturana (whose "autopoiesis" is precisely such a process to generate progressively more and more complex organisms) and by the US mathematician Ben Goertzel (who argued that organisms capable of effectively coupling with other organisms are more likely to survive, and that the coupling process may account for Gould's punctuated equilibrium).

If organisms are composites rather than individuals, then Darwinian evolution can occur much faster and can exhibit sudden jumps to higher forms, and therefore explain two monumental events of life on Earth: how prokaryotes (cells without a nucleus) evolved into eukaryotes (cells that have a nucleus) and the sudden appearance of new species in the fossil record.

The symbiotic creation of species is not such a far-fetched idea. After all, humans can be thought of as collections of organs and viruses co-existing in symbiotic relationships. Generally speaking, the transformation of primitive organisms into more complex ones may be due to the incorporation of other organisms. We know, to start with, that species may also originate by hybridization between existing species, a process that is very common in plants.

Assembling organs in a functionally coherent way is a very difficult task for anybody, including Nature itself, especially if the forces working on it are random; but mixing genomes may be relatively easy. The chemical process that can dramatically alter the genetic code of an organism to incorporate the genetic code of another organism may exploit the very peculiar structure of the DNA double helix and the very peculiar behavior of sex. Both the genetic apparatus and the sexual apparatus seem to be conceived so as to facilitate the fusion of organisms.

While single-organism evolution may explain only gradual and localized changes in skills, the formation of composite structures would certainly result in higher levels of complexity which in turn would result in higher levels of organization.

Unfortunately, we have no idea of how the DNA of a new organism can be synthesized from the DNAs of two organisms, i.e. how a new species can be created by the symbiotic union of two species. The chemical process that allows for the fusion of two codes has not been discovered yet, but may turn out to be a relatively simple "algebra" of the four bases of the DNA.

The Tree Of Life

As geneticists have been rearranging the tree of life based on the DNA or organisms, one thing has become evident: life diverged first into bacteria and archaea, eukaryotes then evolved from archaea but with a little help from bacteria. Somehow eukaryotes acquired genes from bacteria, genes that were critical for their metabolism. This implies that genes are passed not only vertically from generation to generation but also horizontally (or "laterally") from one species to another. This lateral gene transfer could turn out to be the single most important factor of evolution. The more we study their DNA, the more eukaryotes appear only distant relatives to their archaea ancestors, the more they appear the product of a large number of lateral gene transfers. There was probably a time when swapping genes among cells was an ordinary event: by swapping genes, cells would simply trade or share skills with other cells.

Research carried out, among others, by the US biologist Carl Woese is showing that the phylogenetic tree looks more like a web than a tree ("Towards a natural system of organisms: proposal for the domains Archaea, Bacteria, and Eucarya", 1990). By drawing the family tree of today's genes, one should eventually find the genetic content of the common ancestor of all life. Instead, different genes yield different family trees. If they all had forebears in a common ancestor, it must have been a terribly complex being, far from the simple living cell that one expects. It is more likely that some genes were transmitted horizontally (one lineage to another) as well as vertically (one generation to the next one) in the tree. If gene exchanges were common, one can envision a colony of cells as the ancestor of all life and gene exchanges as the main form of early evolution.

Symbiogenesis
In 1909 the Russian botanist Konstantin Merezhkovsky introduced the theory of symbiogenesis. Merezhkovsky viewed living organisms as the result of a combination of two plasms: "mycoplasm" (stuff such as bacteria and fungi) and "amoeboplasm" (basically, eukaryotic cells without a nucleus). Merezhkovsky believed that mycoids were the food of amoeboids, and that one fateful day a mycoid managed to become the nucleus of an ameboid rather than its meal.

A fascination with the wonders of the bacterial world led the US biologist Lynn Margulis (since "The Origin of Mitosing Eukaryotic Cells", 1966) to believe that no other single force has shaped evolution in a more important way. Everything the Earth is today, and everything that we and other living forms do today, is due to conditions that have been created and maintained by bacteria.

Margulis' fundamental thesis is that our bodies are amalgams of several different strains of bacteria. Endosymbiosis of bacteria is responsible for the creation of complex forms of life.

Margulis follows the US biologist Ivan Wallin, who ("Symbionticism and the origin of species", 1927) was the first one to propose that bacteria may represent the fundamental cause of the "origin of species" (Darwin's unsolved mystery) and that the creation of a species may occur via endosymbiosis.

Margulis noted that not all the DNA is contained in the nucleus of the cell. As originally outlined by Wallin, The "mitochondria" are organelles of the cell that function as its "power plants": they convert sugar into energy that the cell can use. Mitochondria have their own DNA, separate from the DNA of the cell. While most DNA is organized as double sets of chromosomes in the nucleus, the DNA of mitochondria stands apart. Margulis believes that the presence of "extra" DNA in the cell is a fossil of an ancient evolutionary event: it attests to the fusion of at least two different kinds of organisms that together formed a "eukaryiotic" cell.

Margulis believes that such symbiotic merger, or "symbiogenesis", has been common in the evolutionary history of life on Earth, and actually accounts for life as we know it today. The ancestors of all life are bacteria. They fused into "protists" (algae, amoebas, etc) which fused into multicellular organisms. Margulis tracks their evolution into plants, animals and fungi.

Margulis emphasizes that the Earth is still dominated by bacteria, which not only account for the vast majority of life, but also maintain the conditions for life on the planet.

All life is either bacteria or descends from bacteria. Life "is" bacteria. Bacteria are also closer to immortality than animals with bodies: cell division generates identical bacterial copies of a bacterial cell. Bacteria can be killed but they do not really die, because countless clones exist of them. The life of a multicellular creature is far more fragile.

Bacteria can also reproduce at amazing rates, compared with "higher" forms of life.

Life can even be viewed as a plan for bacteria to exist forever: bodies are desirable food sources for bacteria, so one could view the evolution of bacteria into such bodies as a plan by bacteria to create food for themselves.

The biosphere is controlled mostly by bacteria, it is, in a sense, "their" environment, not ours. Margulis emphasizes that not only the atmosphere but even the geology of our planet is due to the work of bacteria (mineral deposits have been shaped by the work of bacteria over million of years, or by the reaction with the waste gas of bacteria).

We are allowed to live in it, thanks to the work of bacteria, which maintain the proper balance of chemicals in the air. If all bacteria died, everything would die. It is their world. Every other form of life exists because they exist.

On a smaller scale, if you "fumigated" your body and destroyed all bacteria that live in it, your body would not be able to perform vital functions, such as synthesizing vitamins, and would die.

The mitochondria, which dot all cells of all living beings, are former bacteria, using oxygen to generate energy.

The property of bacteria that intrigues Margulis is that they trade genes, rapidly and easily. DNA is loose inside bacteria's "bodies". Bacteria reproduce by simply splitting their DNA in two. This yields two offspring identical to the parent (same genes). Exchange of genes occurs only when genes are traded among bacteria. Bacterial sex ("conjugation") is about making a new bacterium out of an existing bacterium by adding genes donated by another bacterium.

The new bacterium resulting from the "engrafting" can even change sex, if the "sex" gene is received from the other bacterium (the "sex" gene specifies whether a bacterium is a donor or a receiver).

This process is not really related to our "sex": sex is about two beings making a new being that partially inherits genes from each parent.

When bacteria "create" a new being, they do so by splitting (there is only one parent and the new being is identical to the parent). When bacteria trade genes, a being is changed into another being. Humans do not have either of these processes. I cannot split myself into identical copies of me, and I cannot mutate into another being by absorbing somebody else's genes. (Incidentally, bacteria can also trade genes as plasmids and viruses).

This process of "recombination" occurs even among bacteria of different species. It is as if I could absorb genes from an eagle and turn into a human with wings, and making children who will also be humans with wings. The genetic material of bacteria is extremely flexible and versatile.

Margulis thinks that this is the process that enabled life to evolve rapidly. Scale is crucial: what Margulis realized is the extent to which bacteria rule the planet. They account for a vast portion of the atmosphere and the geology of the planet.

They spread in ancient times and are still spreading today at fantastic speed. Any phenomenon that involves bacteria is involving billions of rapidly moving and mutating beings. Once life was created, once the first bacteria appeared, things happened quickly and on a massive scale. Bacteria spread quickly, thanks to their reproductive efficiency and to their ability to feed on ubiquitous organic compounds.

The first bacteria were "fermenters", feeding on the sugars available on the surface of the planet. They were followed by photosynthesizers: photosynthesis enabled these bacteria to feed on light. Then came bacteria ("cyanobacteria") that could tolerate oxygen, and could therefore feed on water (extract hydrogen atoms from water).

Each new type of bacteria was "polluting" the Earth and therefore changing the environmental conditions for future generations of bacteria. Pollution is an integral part of the evolution of life. The power of bacteria is that their "gene trading" habits made it relatively easy to adapt to whatever new conditions the climate and their own doing were producing.

The history of life is the history of a planet blanketed with rapidly reproducing and rapidly changing beings: the bacteria.

Protoctists were born about 2 billion years ago from the fusion of bacterial cells.

Eukaryotes (living beings whose cells have a nucleus and whose DNA is confined in that nucleus) evolved from those protoctists.

Mitochondria are visible remnants of this process of endosymbiosis.

Experiments by the Korean biologist Kwang Jeon showed that even virulent pathogens can become organelles ("Change of Cellular Pathogens into Required Cell Components", 2006). Margulis concludes that predators can become symbionts, that a deadly infection can become a bodily part.

Margulis extends this paradigm to bodies made of several organs, and suggests that those organs also were accumulated the same way, that they are also due to the fusion with independent organisms by endosymbiosis.

While Darwin was emphasizing competition as the driving process of evolution, Margulis is emphasizing cooperation.

For Margulis life has "free will", and has used it to influence its own evolution. It is not only humans who can affect their environment to direct their own evolution: the whole environment is doing the same. Living beings make decisions all the time and are thus responsible in part for their own evolution, as first speculated by Samuel Butler.

Superbeings

We have not found any evidence of multiple beings integrating in one being, but there is plenty of evidence that individual single-cell organisms sometimes join in creating "collective beings" which are better equipped to survive.

Single-celled bacteria form large colonies in countless ecosystems, particularly visible in seaside locations.

Soil amoebae join together in one huge organism that can react quickly to light and temperature to find food supplies.

Sponges are actually collections of single-celled organisms held together by skeletons of minerals.

These are all examples of how cells are capable of forming communities that live together and live at the same biological "pace". Whereas in a human community we all are independent and interact only occasionally. In such agglomerates of cells every unit is synchronized towards the common goal.

In 1999 the Danish biologist Sune Dano engineered a community of yeast cells that live together as a single organism, driven by collective chemical oscillations.

Among multi-cellular organisms, ants and bees exhibit such a behavior, although the individuals are physically disconnected and communication occurs at a distance through the senses (rather than through chemical contact). Karl Von Frisch, the man who discovered the symbolic dances of the bees, pointed out that the individual is an oxymoron: a bee cannot exist without the rest of the colony. The colony, on the other hand, constitutes a complex and precise self-regulating system that relies on peer-to-peer communication rather than on a dictator imposing order on its subjects. The hive exhibits a personality, the individual is totally anonymous. The way they migrate is even more stunning, as Cecil Johnson described.

The US biologist Deborah Gordon studied ants as a superorganism (the colony as a body, the individuals as cells) and found that the way such a superorganism organizes itself is not too different from the way a brain or an immune system is organized. An ant colony or a beehive seems to have a mind of its own. It has motives and goals, and even exhibits the ability to learn.

After all, what is a body? We tend to think of a body as a set of organs "glued" together, but that is not the case: is blood part of my body? My body cannot exist without blood, but blood is not glued to the other organs. If I make a hole in an artery, blood will pour out. The definition of body is actually quite open. We all believe that ants are quite "intelligent", but we would be reluctant to admit that a single ant shows any intelligence in its random paths of food search and transport. What is intelligent is the colony as a whole. The colony as a whole exhibits stunning coordination and purposeful behavior. The single ant does not compare too well with a human being, but the colony as a whole does. It may be more appropriate to compare our body to the entire ant colony, in which case one notices all the relevant similarities in purposeful behavior: the movement of those ants, taken together, do mimic cognitive, sentient behavior.

A multi-cellular organism is a collection of cells that are synchronized through electrochemical activity. Sponges and amoebae may show how multi-cellular organisms were created from single-cellular organisms. Ants and bees may show that the difference between a multi-cellular organism and a society of organisms resides only in the type of internal communication: they both rely on constituents that are synchronized and the only difference is how those constituents communicate (the dances of the bees as opposed to the chemical reactions of the amoebas).

If this phenomenon cannot help explain evolution as a whole, it can at least shed some light on the transition from mono-cellular to multi-cellular organisms, one of the crucial steps in the evolution of life on this planet.

After all, more than 90% of the cells that make up the human body are not human: they are bacteria (although they weigh a lot less than human cells); and they are vitally important for our survival. There are more than 1000 species of bacteria in the human digestive system alone (and many more in the respiratory system, in the urogenital tract, on the skin, etc). We are a superorganism, or, at least, a walking and thinking ecosystem. All humans share the same genome (99.9% of all genes) but every human is fairly unique when it comes to her or his "microbiome" (even identical twins have wildly different microbiomes). Therefore not only are you a superorganism but, whatever you are, it may be due more to the bacteria that parasite on you than to your own human genes.

Superorganisms

The US philosopher Guy Murchie was perhaps the first to advance the notion that super-organisms are pervasive in nature. The term was introduced in 1876 by the British philosopher Herbert Spencer, and in the 1920s applied to societies of insects by the US myrmecologist William Morton Wheeler.

Inspired by Wheeler, Murchie showed that groups sometimes behave like individual organisms: who runs an ant colony? how do ants decide to move their nest somewhere else? It is the interaction among the individuals: some ants carry eggs and food to the new nest, some ants carry them back, and eventually one of the two competing population prevails (in a sense, "natural selection" decides whether and where the nest moves); bees of a beehive communicate (at least as far as directing their fellow bees to food) with a language which is made of dance steps (including sounds and smells); furthermore, honey bees fan their wings to maintain a constant temperature within the beehive, the same way an organism's parts cooperate to keep the organism within the narrow range of temperature that allows for its survival.

An ant colony or a beehive behaves like an organism with its own mind: a beehive metabolizes, has a cognitive life (makes decisions), acts (it can move, attack) and so forth.

In this scenario, language can be viewed from a different perspective, as the mechanism that allows for the organism to be one.

Murchie envisions the entire Earth as an organism which uses as food the heat of the sun, breathes, metabolizes, and its cognition is made of many tiny parts (organisms) that communicate, exchange energy, interact. All living organisms, along with all the minerals on the surface of the Earth, compose one giant integrated system that, as a whole, controls its behavior so as to survive.

And so do galaxies. After all, we are made of stardust.

Life is inherent in nature. Murchie describes sand dunes, glaciers and fires as living organisms, the life of metals and crystals.

The question is not whether there is life outside our planet, but whether it is possible to have "non-life".

Then Murchie shows that properties of mind are not exclusive to humans. Memory is ubiquitous in nature. For example, energy conservation is a form of memory (an elastic band remembers how much energy was put into stretching it and eventually goes back to the original position). The laws of Physics describe the social life of particles. Electrons obey social laws that we decided are physical laws instead of biological laws thereby granting their behavior a different status from the behavior of bees. But this is an arbitrary decision. Mind can be viewed as a universal aspect of life and energy.

Murchie believes there is one huge mind, the "thinking layer" around the Earth, which corresponds to the "noosphere", a concept introduced by the French paleontologist Pierre Teilhard de Chardin ("Hominization", 1925). Individual "consciousnesses" are absorbed into the superconsciousness of a social group, which is part of a superconsciousness of the world. In Murchie's opinion, the world has a soul, an analogous of the Pythagoreans' "anima mundi" and of the Hindus' "atman".

A Viral Past

Studies on viruses (for example, by the US biologist Luis Villarreal) have also hinted at the possibility that genes could be "acquired" from an external organism, without any need to wait for millions of years of natural selection. A virus is a parasite that comes alive, and replicates, only while it feeds on host cells. This process takes place at the genetic level: the genetic instructions of the virus induce the host cell to manufacture the genes that the virus needs in order to assemble a copy of itself. Thus there is "genetic" contact between the virus and the host cell. Viruses may be the lowest form of life (in fact, most biologists don't even agree that they are forms of life, because they are simpler than living cells), but their fast replication continuously creates new genes, and that process of gene manufacturing takes place inside another organism: the odds that some of those genes get "transferred" permanently to the organism are not negligible. Humans and bacteria share some genes, but those genes are not present in the organisms that should constitute the evolutionary chain from bacteria to humans: how did the intermediary species miss them? The easiest explanation is that somehow the genes of the bacteria "infected" the DNA of humans and became permanent residents of it. Villarreal suspects that the cell nucleus itself of the eukaryotes may have evolved from prokaryotes by, basically, viral infection: the eukaryotic cell might just be a permanently infected prokaryotic cell (the original cell plus its viral invader).

Gaia

Gaia is an idea that originated by the joint work of the British chemist James Lovelock and Lynn Margulis. Lovelock views the entire surface of the Earth, including "inanimate" matter, as a living being (which in 1979 he named "Gaia"), an idea to which the Austrian physicist Fritjof Capra also subscribes. Lovelock and Margulis argued that the rules of life work at both the organism level and at the ecosystem level, and eventually at the level of the entire planet. There is a gigantic cycle that involves the actions and structure of all matter and eventually yields "life" on this planet. The environment (volcanoes, rocks, sea water, sun, rain) is part of life. At the same time life creates the environment that it needs. Life creates the conditions for its own existence.

Capra put it in mathematical form: feedback loops link together living and nonliving matter. The entire planet is a self-organizing network, just like an ecosystem, just like a living system. Living systems are networks interacting with other networks. Organisms are networks of cells. Ecosystems are networks of organisms. Biological systems at all levels are networks. The "web of life" consists of networks.

Murchie, Margulis, Capra and Lovelock view the world as an integrated whole.

The French paleontologist Pierre Teilhard de Chardin (1925) and the Russian geologist Vladimir Vernadsky (1926) even thought that the Earth is developing its own mind, the "noosphere", the aggregation of the cognitive activity of all its living matter. Chardin saw it as the consequence of a natural process of consciousness evolution and as leading to the "Omega Point" of supreme (divine) consciousness. Vernadsky saw it as the consequence of technological progress.

Complexity, Specialization and Cooperation

The British biologist John Maynard-Smith and the Hungarian biologist Eors Szathmary argued that each major transition in evolution turned biological units that were capable of independent replication into biological units that needed other biological units in order to replicate. In other words, each "major transition" seems to produce (or be produced by) cooperation. For example, independently replicating nucleid acids evolved into chromosomes (assemblies of molecules that must replicate together). Also, sexless life was replaced by species that have male and female members, and that can replicate only if a male and a female "cooperate". Ants and bees can only replicate in colonies.

In these major transitions, sets of identical biological units were replaced by sets of specialized units that needed to cooperate in order to survive and replicate.

This also opens a window on the history of socialization, or cooperative behavior. Far from being a recent invention, socialization arose when specialization arose. Originally, one can envision a world of multifunctional self-sufficient biological entities. When these evolved into specialized entities, the need for them to socialize was born. Division of labor among a group of specialists is more effective than a multifunctional non-specialist but only if the specialists cooperate. And thus the multifunctional cell led to cellular organization and eventually to bodies with specialized limbs and organs that eventually led to societies of specialists (ants, bees, humans). Altruism, or at least division of labor and cooperation, appeared very early in the history of life, as soon as molecules were enclosed within membranes.

After all, cooperation was inherent in Mendel's laws: a gene's chances of surviving in future generations depends on the success of the cell that hosts that gene, a success that depends on the success of all the other genes that determine the life of that cell. Hence a gene has a vested interest in "cooperating" with the other genes. The cell would not survive if its genes did not form an efficient society.

Synergy

The US biologist Peter Corning believes that "synergy" is pervasive in the universe at all levels of organization, and plays a role in producing "variation", the phenomenon that makes natural selection possible. Corning argues that traditional Darwinism cannot explain complexity (on large scales) precisely because its emphasis is on competition and not cooperation. Genetic mutations per se would not be enough to explain the complexity we find in nature. Corning instead focuses on the behavior of living beings, that are capable of learning (the "Baldwin effect") and are capable of modifying the environment (as Waddington and Mayr pointed out). Living beings are more than mere "vehicles" for genes to live forever. Living beings actively participate in determining their own evolutionary future by 1. Continuously reshaping the environment that will "select" their evolution (the use of tools is pervasive among living beings) and 2. Learning behavior that is not in their genes and passing it on to the next generation (learning is also pervasive among living beings). In other words, living systems shape the very environment that drives their evolution. He goes as far as to claim that humans, the most active living systems, have "invented themselves" by creating the environment that they wanted.

Behavior shapes evolution. In particular, humans adopted "group-based behavioral strategies", i.e. social organization. He emphasizes the importance of tools to shape our behavior, in particular a dietary shift from vegetables to meat. That, in his opinion, caused subsequent anatomical developments of the hominids. Climate change caused behavioral changes, and they caused anatomical changes. Language is also a by-product of behavioral changes that, in turn, fostered anatomical changes.

Corning emphasizes the importance of the transfer of culture from one generation to the next one. Culture, in turn, helped create novel forms of synergy, or, in other words, higher levels in the hierarchy. We are still creating new forms of synergies.

Corning thinks that two quantities need to be added to Monod's "chance" and "necessity": teleonomy and selection (selection was implied in Monod's theory although not in the title of his book). Teleonomy is a property that living systems exhibit: their structure and function has a purpose and is directed towards a goal. This property is a consequence of the living system's evolutionary history. Teleonomy is coded in the genome of the living system. The genotype determines the behavior of the phenotype, but the phenotype in turn helps to create the selection pressures that will determine the evolution of the genotype. Teleonomy has an impact on evolution because it is a form of downward causation: the behavior of the whole creates the selection pressures that cause the evolution of its "parts" (all the way down to the genes themselves).

Nature is organized in a hierarchy of hierarchies. At each level of a hierarchy different "synergies" are at work that create the upper level. Nature's creativity lies in the combination of parts to create wholes that are more than the sum of their parts. The universe is still inventing itself and we are not just spectators but co-protagonists.

Corning, therefore, believes in "synergistic selection", which is Darwinian selection at the level of complex systems: the differential survival of wholes that leads to the emergence of higher-level wholes whose purpose transcends the purpose of their constituent parts. These wholes in turn become agents of selection for both themselves and others. Corning's "Neo-Lamarckian Selection" is not in opposition to Darwinian selection but complements it.

Just like Robert Wright's "nonzero sum game", Corning's theory is fundamentally a theory that says cooperation is important in human evolution.. The difference between the two is that Wright believes in an inevitable destiny towards greater complexity and progress driven by "non-zero sumness", i.e. by a fundamental mathematical law that rewards cooperation, whereas Corning believes that we are free agents of our own future. Corning points out that for every giant step ahead the human race has stumbled into an equally impressive step backwards. So the direction of history is not clear at all.

Sex And Death

Bacteria reproduce by replication and mutate by conjugation. Mitosis ("the dance of chromosomes") is the process by which eukaryotic cells reproduce: the DNA of the new being is a combination of the DNA of the two parents. In eukaryotes the DNA is not just a string: genes are organized in chromosomes (a minimum of two, humans have 46).

Prokaryotes are wildly different from bacteria. But how did this striking difference between bacteria and their descendants come to be?

Mitosis is truly responsible for the origin of species. Before mitosis, bacteria were freely exchanging genes: the concept of "species" as we know them today did not exist, as any bacterium could mutate into a novel "species" at any time. Bacteria do not have true species.

On the other hand, multicellular beings cannot trade genes. Therefore they cannot mutate into anything else, and their offspring belongs to the same species (because both parents must be of the same species in order to interbreed) and such offspring inherits genes of the parents. Genes remain within the same family, the "species". Any multicellular being is a member of a species: it is an obvious fact, but a quite striking one. It is one of nature's whims. At the beginning of life on Earth, a new bacterium could be just about any combination of available DNA. Later in evolution, a new individual had to be a member of a species.

It may not be a coincidence that death was invented with multicellular sexual beings. They age and die, whereas bacteria did not.

Why did sexless and immortal bacteria evolve into beings that have sex and die?

Bacteria have only one sex, they can mutate (change their DNA), they can interbreed with any bacteria, they don't make children, and they never age or die. Animals that evolved from them have two sexes, they cannot mutate (cannot change their DNA), they can only inbreed within their species, they make children and they age and die. (Last but not least, the DNA of animals is organized and inherited in units called chromosomes, a detail that may turn out to be crucial to explain all of the above discrepancies).

Margulis argues that "death was the first sexually-transmitted disease". Once animals started aging and dying (once death had been programmed into their DNA), their offspring inherited the same disease.

Margulis' hypothesis is that, once upon a time, "eating and mating were the same". Cannibal unicellular beings may have merged into multicellular beings. The evolutionary advantages of this behavior may have led to sexual beings.

But the genders are exactly two, and each member of a gender has the same sexual organs. How did that happen?

Guy Murchie believes that death provides an evolutionary advantage: immortal beings that simply split would be immutable and easy prey to environmental changes. Death allows for regeneration of the race and for creation of new species. Death is a tool for change and progress. It is not a coincidence that the odds of immortality increases as creatures get more elementary.

Notwithstanding these cunning speculations, sexually-reproducing species are a bit of a mystery, and so is death, that came with sex.

The Origin of Selection

According to modern synthesis, the genetic makeup of a population is altered through natural selection (the interaction between the individuals of the population and their environment). Darwin's approach to the problem implied that natural selection mainly acts on the individual (precisely, it causes differences in phenotype among individuals within a population), although he explicitly recognized three levels of natural selection: individual selection, kin selection, group selection. Several biologists have argued that selection might act at a number of different levels, loosely corresponding to a hierarchy of biological organization: genes, individuals, kin, groups, populations, and species. Ultimately, what changes is species, but that is the effect of a process of natural selection that may act at any of these levels and then cause that visible effect on species.

Evolutionary theory is based upon the idea that species evolve and their evolution is driven by natural selection, but it is not clear what exactly evolves and what natural selection acts upon. Nature is organized in a hierarchy: genes are located on chromosomes, chromosomes are located in cells, cells make up organs which make up organisms which make up species which make up populations which make up ecosystems: at what level does selection act? One may view the genes as the units that must change to generate evolution. Or one may view ecosystems as made of co-evolving species that would not evolve the same way by themselves. And so forth.

Gould supports a hierarchical model that views selection as acting simultaneously at a variety of levels in a genealogical sequence of gene, organism, population and species.

David Sloan Wilson views nature organized in a structural hierarchy, and selection acting at each level of the hierarchy, but which levels matter more depend on the species. In the case of humans and other species, the group (hive, herd, clan, tribe, nation) was one of the most relevant levels.

The German biophysicist Bernd-Olaf Kuppers thinks that natural selection applies to the molecular level.

The US biologist Richard Lewontin thinks that all entities that exhibit inheritable variance in fitness (from pre-biotic molecules to whole populations) are units of selection. The US philosopher Robert Brandon thinks that the biosphere is hierarchically arranged and, in agreement with Lewontin, natural selection applies to all levels of the hierarchy.

For the US zoologist Terrell Hamilton, the individual is the unit of natural selection. He separates selection, adaptation and evolution: natural selection results in differential reproduction, therefore, in adaptation of populations, therefore, in evolutionary change. Correspondingly, the individual is the unit of natural selection, gene substitution is the elementary process of adaptation, and the species is the main unit of evolution.

Alfred Russell Wallace, co-inventor of evolution theory with Darwin, thought that selection acts on populations as well as individuals. Selection at the level of populations occurs when a group of individuals produces more groups than other groups.

The British biologist Richard Dawkins popularized "gene selectionism", according to which the genes compete and are responsible for evolution.

Finally, Ernst Mayr thinks that genes cannot be treated as separate, individual units, that their interaction is not negligible. The units of evolution and natural selection are not individual genes but groups of genes tied together into balanced adaptive systems. Natural selection favors phenotypes, not genes or genotypes. Ultimately, species are the units of evolution. After all, speciation is the method by which evolution advances.

The US chemist Jeffrey Wicken thinks that the most general entities subject to natural selection are neither genes nor populations but information patterns of thermodynamic flows, such as ecosystems and socioeconomic systems. Natural selection is not an external force, but an internal process such that macromolecules are accrued in proportion to their usefulness for the efficiency of the global system.

The US biologist William Wimsatt grounds the notion of selection around the notion of "additive variance". This quantity determines the rate of evolution. Variance in fitness is totally additive when the increase of fitness in a genotype is a linear function of the number of genes of a given type that are present in it. If variance in fitness at a given level is totally additive, then this is the highest level at which selection operates. The entities at that level are composed of units of selection, and there are no higher-level units of selections.

Gene Selectionism
Richard Dawkins and the British philosopher Helena Cronin argue that genes rather than organisms (as Darwin held) are the primary units of natural selection.

Dawkins essentially built on the work of the US biologist George Williams. Williams thought that genes encouraging altruism would quickly become extinguished, and therefore genes must be "selfish" in nature. Every trait serves some kind of self-interest. Genes that serve that self-interest are more likely to survive (because their vehicles are more likely to survive) and multiply. Thus the corresponding traits are more likely to become widespread among future generations.

Dawkins introduced whole new methods of thinking about life, behavior and evolution. Firstly, Dawkins argued that the gene is the fundamental unit of evolution: genes drive evolution and genes drive behavior.

Darwin's assumption that natural selection favors those individuals best fitted to survive and reproduce can then be restated as: natural selection favors those genes that replicate through many generations. The level at which selection occurs is not that of the individual organism, but that of particular stretches of genetic material. Organisms are merely the means that genes use to perpetuate copies of themselves. The universe is dominated by stable structures, and one particular stable structure is a molecule that makes copies of itself.

A "replicator" is an entity that copies itself, such as genes. A "vehicle" is the organism that carries the replicator in its cells and whose differential survival and reproduction results in the differential spread of the replicator. Dawkins thinks that the superiority of replicators is obvious. A replicator serves as a repository of information (about the organism but also, indirectly, about the environment) that is preserved over time and spread over space. Replicators are immanent entities: they exist virtually forever. Vehicles, on the other hand, are merely "tests" of how good that information is. And, of course, vehicles are also the machine used by replicators to copy themselves.

The US philosopher David Hull offered a slight variation on Dawkins' theme. Hull distinguishes replicators (units that reproduce their structure directly) from "interactors" (entities that interact directly with their environment). Darwin's theory of evolution through natural selection thus reads: differences in the interactions of interactors with their environment result in differential reproduction of replicators. The difference between Hull's "interactors" and Dawkins's "vehicles" is not trivial: genes are both replicators and interactors (they have a physical structure that interacts with an environment), and some interactors are also replicators (the paramecium that splits in two).

However, the general scheme remains the same. Natural selection is about the differential survival of replicators. Genes can be replicators whereas multicellular organisms, groups and other levels of the hierarchy can only be vehicles/interactors.

In other words, what survives is not my body but my genes. It is not bodies that replicate when children are made: it is genes that replicate in the children. Therefore, natural selection can't be about bodies, it must be about genes. Bodies are in a lose/lose situation, as they will disappear anyway. But genes do have a chance to survive (by copying themselves into a new body).

Of course, this doesn't mean that genes "are" eternal. Genes are perpetuated insofar as they yield phenotypes that have selective advantages over competing phenotypes. They have a chance of being eternal, but that depends on how good they are at creating competitive organisms.

An organism is a mere gene-transporting device: its primary function is not even to reproduce itself, albeit to reproduce genes. The mind itself is engineered to perpetuate DNA. The brain is a machine whose goal is to maximize fitness in its environment.

From the point of view of a gene, any organism carrying it is an equivalent reproductive source. In many cases siblings are more closely related (genetically speaking) than parents and offspring. Adaptation is for the good of the replicator. Therefore, it is not surprising that sometimes organisms sacrifice themselves for improving their kin's survival. Kin selection is part of a gene reproduction strategy.

"I" am not the subject: I am the object. My genes are the subject. I am but a product of my genes. Genes represent a higher force than my will, a force that has been acting for millions of years, compared to the few decades that my will be performing. Genes tell me what to will. Genes tell me how to interact with other people who are the product of other genes, i.e. they tell me which genes to interact with. Genes tell me what food I should eat and what dangers I should avoid. Whether there is a conscious entity in my genes or not, it is "them" that drive my existence. It is "them" who want me to reproduce: I will be dead soon, but they will still be somewhat alive in my relatives. My family is not going to be extinguished any time soon. I will be a mere step, soon forgotten and useless, in their process of reproduction, of survival, of progress.

Genes want to live forever.

The Altruistic Gene

The British zoologist Mark Ridley makes a distinction between the macroscopic effects and the microscopic causes of animal behavior.

The puzzling feature of the animal world is that animals often help each other, and sometimes some individuals would sacrifice their lives to save others. This would not make any sense if the goal were merely for the body to survive.

Altruism was explained by Richard Dawkins with the idea that evolution applied to genes, not to bodies. Bodies are the vehicles that genes use to attain everlasting life. Bodies are disposable. Genes are not used by organisms, genes use organisms. I am nothing but a machine invented by a bunch of genes to maximize their chances (not mine) to survive. I will die. But if I am fit and make children, my genes will survive me. And if my children are fit, they will die but those genes will continue to exist in other bodies, generation after generation. It's the genes, not the organisms. Darwin's idea of competition among individuals for survival must be slightly modified: it is not individuals that compete, it is genes. In order to maximize its chances of survival, a gene would cause one of its bodies (one of the bodies that contain that gene) to help its "kin" (bodies with the same gene). The macroscopic effect would be cooperation among organisms, while at the microscopic level that cooperation is truly an attempt by the gene to outsmart other genes, i.e. it is competition of the most cynical kind.

You have to think like a gene, not like a body. If you are a gene, you have no problem sacrificing some of your bodies to save some others. Your ultimate goal is to survive (you are the gene) and you can use any of those bodies as vehicles to continue your journey through time. Altruism makes as much sense as selfishness in the classical Darwinian theory, as long as you look at the micro-world, not just at the animal kingdom (the macro-world) as we (bodies) see it.

In mathematical terms, sex provides a way for a gene to participate in a lottery a number of times: each body is a participant in the lottery of survival. The more bodies, the more chances to win the lottery.

This is a special lottery, though. Winning this lottery entails some work (creating and maneuvering the organism) and this work must be done jointly with other genes. Sex is the process by which a gene is chosen to work in a body together with other genes.

In each offspring the gene is working with a different set of genes. Each offspring is a combination of genes. Some of those combinations will prevail, i.e. they will generate an organism that is capable of surviving in the environment. The gene has a vested interest in that as many as possible of those offspring survive. If you are one of those offspring, you think that it is all about you. But, in reality, it is all about the genes that are inside you, and that you share with your siblings (and some with your cousins, and some with your entire tribe, and some with the entire human kind).

If you are a gene shared by my brother and me, it makes perfect sense that i give my life to save my brother's children. I am not jeopardizing my chances of survival: i am maximizing your chances of survival.

Matt Ridley sides with Dawkins in thinking that the gene is the unit of selection and in believing that genes are selfish; but Ridley shows that it is in their interest to form alliances, because that may increase the chances of survival for their genetic pool. Cooperation is actually a recurring theme at all levels of the biological world, from cells to species. Ridley explains cooperation among organisms of different species by using game theory: whenever the mathematics of benefits outweighs the mathematics of competition, organisms tend to be cooperative. Therefore, Ridley believes that social behavior, such as cooperation, trade, religion, is a direct consequence of evolution.

Selfish Altruism

Altruism could be a simple outcome of a cost/benefit analysis that begins at home and continues in the world at large.

Altruism does not seem to be innate, not even among siblings. Children are selfish. It takes years to teach children to be "nice" to other family members. If i have a candy and my brother has a candy, i want his candy and i don't want to give him my candy.

However, i quickly learn that my parents will punish me if i steal his candy but will reward me if i give him my candy. Therefore at some point i become a "good kid" who does not steal my brother's candy and instead offers him my candy. The long-term goal of gaining my parents' affection, protection and trust prevails over the short-term goal of getting as many candies as i can.

Parents teach us to be nice to our siblings because parents care for all their children. They blackmail us into being nice to the other siblings by threatening punishments and promising rewards. We are naturally inclined to be altruistic to our brothers and sisters: our parents are, and our parents instill that value in all their children.

Society at large does the same for all individuals: be nice to others, even complete strangers, and society will reward you with protection and respect.

The US social scientist Howard Margolis argued in favor of a compromise between the two views. He speculated that we have two selves, one selfish and one altruistic, and our behavior is the outcome of a rational, Darwinian strategy on how to allocate resources between those two selves.

Sociobiology

In the 1970s the US biologist Edward-Osborne Wilson popularized "Sociobiology", the discipline that studies the biological basis of social behavior. Wilson's tenet was simply a generalization of William Hamilton's ideas: that the social behavior of animals and humans can be explained from the viewpoint of evolution, that human behavior is largely determined by our genome.

Sociobiology, for example, should be able to explain why murder is almost exclusively a male phenomenon and it peaks at the age of 25. Sociobiology aims at tracing the evolution of humans and their habits, from sex to language.

The central tenet of Sociobiology is that all aspects of human culture and behavior are coded in the genes and have been molded by natural selection. Wilson is after a biological explanation for everything: religion, ethics, and ultimately for the history of humankind. His program is to identify universals in human societies. Ultimately, the aim is to define human nature. Wilson believes that universals are coded in the human genotype; and, like everything else coded there, they have been selected by evolution.

Wilson attempted a unified theory of Biology and social sciences, from genes to mind to culture. The underlying theme of his findings is a strong coupling between genetic and cultural evolution. They proceed together, in parallel and intertwined.

Wilson defines culture as the product of the interaction of all the mental and physical artifacts of a population. Culture is not unique to humans. What is unique about human culture is that it is a form of "euculture", which involves "reification" (the construction of concepts and the continuous re-categorization of the world, including the ability to symbolize), besides teaching, imitation and learning (which are present in many other animals).

On these foundations, Wilson reconstructed the genetic history of our mind.

A culture expresses itself through its "culturgens" (behaviors and artifacts). These are the equivalent of genes. These are the basic units of inheritance in cultural evolution. Each individual is genetically endowed with epigenetic rules to process culturgens. Such rules can be said to assemble the mind of an individual. They include sensory filters and cognitive faculties, all of them determined genetically. And, ultimately, these rules affect the probability of transmitting a culturgen as opposed to another.

Epigenesis is the process of interaction between genes and the environment during development. Epigenetic rules affect both primary functions such as hearing and secondary functions such as mother-infant bonding and incest avoidance.

One or more culturgens are favored by the epigenetic rules. Eucultural species such as humans evolve towards a type of cultural transmission in which a dual shift occurs in time: change in the epigenetic rules due to shifts in the genes frequency and change in culturgen frequencies due to the epigenetic (gene-culture co-evolution). The two shifts exert a mutual influence.

The epigenetic rules exhibit genetic variation, thereby contributing to the variance of cognitive traits within a population. The fitness of the individuals differs depending on their minds' behaviors. Therefore the population as a whole tends to shift towards the most efficient epigenetic rules.

The general model of Wilson is one in which the offspring learn to "socialize" from both their age peers and their parents. They evaluate the culturgens and assimilate them depending on their epigenetic rules; and then use the outcome to exploit the environment.

Wilson's ambitious program is to unify all disciplines of human knowledge (from religion to art) in one discipline ("consilience"), which would be, fundamentally, the study of how the human mind evolved. He believes that all other disciplines could be reduced to this discipline, and therefore what they study are but particular aspects of the evolution of the mind and, ultimately, of its genetic programming.

Culture is therefore a product of biology. Culture is social behavior.

Evolutionary Psychology

Evolutionary psychology was pioneered by the US anthropologist John Tooby and the US psychologist Leda Cosmides. They believed that culture shapes human behavior notwithstanding biological pressures, and therefore disagreed with purely biological explanations of social behavior such as Wilson's and Ridley's.

Evolutionary Psychology basically investigates the biological origins of human behavior. For example, it studies the different patterns of behavior of males and females based on their roles in sexual reproduction (the male's only investment is in spreading his sperm as widely as possible, whereas the female's investment is much bigger and involves both bearing and nurturing the offspring). Natural selection has molded the brains of men and women in different ways as a result of their different reproductive goals.

Evolutionary Psychology rests on the seminal work of a number of biologists who dealt with the genetic foundations of high-level behavior, starting with William Hamilton ("The Genetic Evolution of Social Behavior", 1963). The British geneticist Angus Bateman had already suggested ("Intra-sexual selection in Drosophila", 1948) that natural selection had determined different male and female behaviors. The US biologist George Williams formalized this idea in a different way: the "sacrifice" required for reproduction is different for the female and the male. In 1972 Robert Trivers replaced "sacrifice" with (parental) "investment": the investment required for reproduction (to increase the chances of survival of the offspring) is different between a male and a female, and that accounts for different attitudes towards the other sex and the offspring itself.

These biologists applied Darwinian thinking to the social behavior of animals. These studies, once applied to humans, laid the foundations for Evolutionary Psychology, basically a more scientific way to study human behavior than Psychiatry. In fact, Evolutionary Psychology is not about human behavior: it is about human nature (which determines human behavior).

Most of an organism's behavior is mechanical, instinctive, although it makes a lot of sense: all the "thinking" has already been done by natural selection and summarized in its DNA. Genes determine behavior that has been found to be rational over thousands of generations of testing. If it were not rational, those genes would not have survived, and that behavior would not exist.

Evolutionary Psychology introduced a new kind of "unconscious": the control that comes from the genes.

The Amoral Animal

Thus the US historian Robert Wright replaces Freud's subconscious with Darwin's natural selection as the engine of all adult behavior.

Darwin himself had already realized that animals are subject to one kind of pressure that comes from members of their own species: sexual selection. Males have to compete in order to mate with a female. Females get to choose which male they mate with. Males seem to be indiscriminate in their sexual appetite, whereas females seem to be very discriminating. This simple asymmetry of behavior explains many traits that would not be easy to explain with standard Darwinian theory (for example, why some animals have very colored traits, and proudly display them, thus helping their predators spot them). But sexual selection often prevails: males who were not equipped to compete against other males (e.g., bulls with no horns) and to attract females (e.g., peacocks with small tails) were excluded from sex, and their traits are thus extinct.

Darwin did not explain where sexual selection comes from, though. George Williams found the answer. It comes from a simple physical fact: women can reproduce only about once a year, whereas men can reproduce every day of the year (if they find a woman willing to, of course). For a woman the main "investment" to reproduction is giving birth and nurturing the baby, a lengthy and complex consequence of a few minutes of sex. For a man the main investment is just those few minutes of sex. Thus the different sexual behavior.

Human nature does not come out looking too good. Human nature is merely a machine that has been fine-tuned over millions of years to maximize a mathematical equation (that of survival of our genes).

Wright showed that morality is simply the set of rules that increase the odds to pass one's genes to the next generation. Morality is mere convenience. To become moral animals, Wright claims, humans must first realize how thoroughly amoral they are.

The Mind As A Sexual Organ

The US evolutionary psychologist Geoffrey Miller believes that natural selection per se cannot account for the sophistication of the human mind. He thinks that an extra force must have been at work. That force is the combined effect of sexual choices that our ancestors made. They helped design us as we are today.

Miller views the human mind not as a problem solver, but as a "sexual ornament". Miller points to the fact that the human brain's creative intelligence must exist for a purpose, and that purpose is not obvious. Survival in the environment does not quite require the sophistication of

Einstein's science or Michelangelo's paintings or Beethoven's symphonies. On the other hand, these are precisely the kind of things that the human brain does a lot better than other animal brains.

The human brain is much more powerful than it needs to be.

Miller explains the emergence of art, science and philosophy by thinking not in terms of survival benefits but in terms of reproductive benefits. Miller basically separates (as Darwin originally did) natural selection (competition for survival) and sexual selection (competition for reproduction). Then Miller argues that sexual selection is much more efficient and intelligent, because it is not driven by random environmental events but by a deliberate strategy to improve the "genetic quality of the offspring".

Sexual selection is as intelligent as we are, whereas natural selection (from the viewpoint of human self-interest) is hardly intelligent at all (it does not intentionally reward humans over other species, or one individual over other individuals).

Sexual selection is a form of positive feedback (as Ronald Fisher had already showed in 1930), the kind of process that can explain the explosive growth of the human brain.

Miller argues that a fundamental function of the human mind is to display one's fitness to the other sex. As Darwin originally proposed, sexual selection originated from two parallel and interacting processes: men compete for women, and women choose men. Thus bulls have horns (to fight other bulls) and peacocks have tails (to attract women): these are organs that serve no other function. They evolved because of sexual selection. As Ronald Fisher had already showed, sexual selection can rapidly lead to evolution of sexually-relevant traits in animals: as females get pickier, they pick more attractive males, thus making children that are more attractive and who will therefore make more children. Evolution favors both pickier females and more attractive males. Thus the ornaments of several male animals evolved. This cycle continues (spirals up) until the ornaments become counterproductive to the other selection process, natural selection.

Men had to be accepted by women in order to make children. We are the descendants of those who were "sexually selected". Miller believes that sexual selection was based on activities such as painting, singing and dancing (which, in turn, explains why humans paint, sing and dance). Miller shows that each of these activities turns out to be a good indicator of physical and mental fitness, that women recognize, evaluate and reward with sex. Birds do the same when they sing complex melodies, and fruit flies do the same when they perform complex dances. These are all activities that appear to serve no survival purpose but appear to contribute to reproductive success. Males need to advertise their genes, and this need drives innovation.

Artistic activities developed because they contributed to sexual selection. When language appeared, it allowed thought itself to be used for sexual selection. The growing importance of thought for sexual selection drove, in turn, the evolution of language. Sexual selection has slowly shifted from body to mind.

Sexiness

In the 1980s the US biologist Russell Lande had already argued that sexual selection is a form of positive feedback, whereby features that attract the other sex tend to increase and, at the same time, the preference for those features in the other sex also tends to increase. The offspring of the mating caused by those preferences is likely to inherit the same preference and the loop continues at ever increasing speed. The result is an exponential increase in the "sexy" feature, whether a tail or a sound. Basically, sexual attraction is an inherited mistake.

In 1993 the US biologist Randy Thornhill showed that several of the most common attributes of beauty could constitute a good indicator of a healthy body. What we perceive as "beautiful" in a woman or man turns out to be simply a good immune system. Evolution taught us how to tell from the outside how good a person's immune system is.

His colleague, Austrian ethologist Karl Grammer, showed that even a person's smell could be a good indicator of that person's immunological strength.

Cooking

The British anthropologist Richard Wrangham argues that the discovery of fire, about 1.8 million years ago, caused a major evolutionary change corresponding to the emergence of Homo Erectus. Humans started cooking, the first and (so far) only animal to do so, which resulted in the ability to extract more calories to support big brains. This fueled the evolution of both the body (that could survive with smaller jaws, teeth and guts) and the brain (that could grow virtually at will). The advent of cooking also had an impact on the evolution of society. For example, it freed men and women from the chore of chewing food that tends to require much more time in other primates. Men and women had more time during the day to invest in other activities. According to Wrangham, cooking is not only a product of human evolution, but also the enabling factor of it.

The French anthropologist Claude Levi-Strauss had already emphasized that cooking prompted the transition from "nature" to "culture" because it separated (once and for all) humans from other animals and it heralded the conquest of nature by humans.

However, the only part of the story that has solid foundations is that fire significantly changed the daily life and the chances of survival for early humans. It provided (besides cooked meals): light in the dark, warm temperatures in winter, a way to dry clothes after a rain, protection from bigger mammals, a signaling system, etc. Each of these factors laid the foundations for major changes in human civilization: longer periods of activity, survival in cold climates, safer dwellings, long-distance communications and, sure, a more efficient diet.

A Woman's World

The US surgeon Leonard Shlain realized that we tend to examine the evolution of humans from the male perspective. We end up with a lot of mysteries as to why humans did what they did. He looked at it from the female perspective, and came up with a lot of interesting solutions to those mysteries.

Shlain's line of argument is fundamentally that, because of the explosion in size of the human brain, women a) did not want to have children (they would die by the thousands while giving birth) and b) they desperately needed iron (a key ingredient of brains). Both factors caused women to develop strategies of mating that forced men to change their habits. The danger of giving birth led women to decouple sex and reproduction. The search for iron led females to favor men who could hunt and bring meat, a form of concentrated iron.

"Males tended to be what females wanted them to be".

The ultimate paradox of the human female is that she needs a lot of iron, but she loses a lot of it when she menstruates. No other female loses so much blood (and the ones who menstruate are a small minority anyway). What purpose can menstruating possibly serve? And why would menstruation be synchronized with the motion of the moon, a very distant object whose only visible effect on the Earth is the tide?

Shlain travels back in time to the day when a baby with a large-size brain was born to a woman whose pelvis was too small for it. Homo Sapiens posed a danger to the woman who gave birth to it. One of the leading causes of death among human females was and still is birth (where medical facilities are still primitive). This is clearly another paradox of nature: why should a mother be punished with death for giving birth to a child? This large-sized brain was the product of a random mutation but it must have turned out to be able to cope with the challenges of the environment better than the smaller brains (that presumably caused a smaller number of deaths among mothers) if today there are six billions of such large-size brains on this planet and they pretty much rule the planet. Shlain believes that evolution took care of the paradox not by modifying the body of the human female (the "natural" thing one would expect to happen is women with larger vaginas) but by modifying her behavior.

The females of the vast majority of species clearly signal to the males that they are ovulating and that therefore they welcome sexual attentions. Her signals (the "estrus"), that range from vocal sounds to bodily changes, arouse the males, and, therefore, increase the chances of having sex when it makes more sense for the purpose of reproduction (which is the only purpose known to almost all species expect for humans). The human female, though, is clueless about her period of ovulation: there is no physical evidence that she is ovulating. Neither is she aware of it, nor can the male be. No wonder that today sex and reproduction are uncoupled among humans to an extent unmatched by any other species. There are many more "oddities": the human female experiences orgasm, which is unknown to other species in that intense form; the human female has sex facing her partner; the human female has sex even "on top"; the human female experiences menopause (she stops ovulating altogether, i.e. of being able of making babies) relatively early in her life. Notably, the female of every other species copulates when it ovulates. The human female is the only one that decouples the two activities.

The larger-size brain of Homo Sapiens (mostly due to the appearance of the neocortex) was capable of new cognitive tasks. Shlain believes that it was women who first exploited the neocortex's new capability for reasoning. The human female realized something that no other species seem to have realized: the connection between sex and pregnancy. In women this was the connection between sex and the painful, life-threatening event of giving birth. Once women realized the connection (that each sexual intercourse could cause their death within nine months), they became less interested in sex than their counterparts in all other species. Basically, the human female became the first female capable of saying "no". Shlain believes that it was the woman, not the male, to undergo a major psychological and physical transformation because it was her life that was on the line.

Shlain believes that the human male reacted by developing strategies based on what "she" wanted and was willing to trade with sex. It turned out that the human female needed iron; lots of it, more than the previous hominids had been able to get from vegetables. Hence the human male became a hunter of animals and Homo Sapiens became an omnivore. Shlain believes the whole bloodthirsty nature of humans (from duels to genocides) is due to that original cause.

Males of other species are not interested in having sex with women who are not ovulating. Only the human male is constantly seeking sexual intercourse. The human male is also the only one to engage in convoluted forms of sex. Shlain believes that this sexual urge serves a purpose whose beneficiary is the female: because hunting was so dangerous and unnatural for humans, the man had to be highly motivated to risk his life hunting an animal on behalf of a woman. Nature motivated man by equipping him with an obsession for sex. Because he was constantly in need of sex, the human male was willing to go and hunt bigger and stronger preys despite his physical limitations.

As further evidence that humans were originally vegetarian, Shlain mentions a weak digestive system and an immune system that does not protect against the germs of rotting meat: no other animal has to cook meat before eating it. Cooking requires fire. Shlain hints at the possibility that fire itself was invented to take care of women's passion for meat.

Another oddity of Homo Sapiens is that men reach their sexual peak when they are young (which makes sense because it's when they are producing the largest amount of sperm), whereas a woman's libido peaks in her 30s, when she is actually approaching menopause. It makes no sense that women want to have sex at an age when it would be wiser not to have children. Even more puzzling is that in all societies there seems to be a general preference for younger women among men and for older men among women. The net result is that no other species masturbates so often as Homo Sapiens. Shlain has a simple explanation. The greatest threat to a young woman's life was an early pregnancy, because her body was not equipped for having babies with large brains. The greatest threat to a young man's life was hunting, because his body was not equipped for fighting with large, fast, strong animals. A reduced libido helped young women abstain until their pelvis grew larger. An increased libido prompted young men to go hunting.

Shlain believes that the orgasm of the human female, which is a unique phenomenon because of its intensity and duration, is nature's way to make a woman commit to the irrational act of having sex, an act that may cost her life when the baby is born. Since sex can be very pleasant, women have a motivation to do something that otherwise they would never want to do, thereby decreeing the extinction of the human race. (Shlain believes that this was the original purpose of circumcision, as circumcised men take longer to have an orgasm, thus giving women time to achieve their own).

(One could add that the asynchronous phases of sexuality between men and women may also explain cosmetics and fashion, both of which help women increase their appeal at an age when their physical appearance declines at the same time that their libido increases).

Shlain believes that female sexual behavior is also responsible for the evolution of the concept of time, and for memory in general. Shlain believes that humans are the only species that lives in a four-dimensional world. Plants live in a world with no space and no time dimensions. Primitive organisms capable of moving lived in a mono-dimensional space. Larger animals live in a three-dimensional space. But Shlain thinks that only humans are aware of Time because only humans can replay the past in their memories. The evolution of memory and the sense of time was vital for men to improve their ability to plan that was vital to become a successful hunter in a world of bigger, stronger and faster animals. Shlain thinks that this feature was, again, driven by female sexuality: the monthly periodicity of menstruations serving as a clock to "measure" time.

Synchronizing them with the moon helped make them even more useful as a clock. (Homo Sapiens is among the few animals whose visual system and whose habitat make it easy to see the moon). Thus Shlain believes that menstruations ultimately helped humans evolve the temporal dimension and the cognitive ability of anticipating the future. The temporal dimension made them more efficient hunters.

Alas, the temporal dimension soon led to the realization of our mortality.

Another side-effect of the temporal dimension, according to Shlain, is that man became obsessed with paternity. As he realized what she had already realized, that a baby is the outcome of a sexual intercourse, he became paranoid about knowing that her babies were his. Thus the escalation of laws meant to limit a woman's sexuality that make it easy for a man to know that her children are his. The institution of patriarchy followed shortly thereafter.

Shlain believes that language too had a sexual origin. As females became more aware of Time and of the implications of a pregnancy, language basically replaced hunting as a way for females to assess how good a mate each male would be. Shlain believes that the original purpose of language survives in today's habits, when men romance women while women mostly listen.

Human Contradictions

All in all, the sexual behavior of humans is relatively easy to explain. Much harder is to explain our attitude towards death. Since prehistoric times, humans bury the dead. Sometimes the deceased is buried with some of his/her belongings. Sometimes the children of the dead worship the dead. In fact, worship of the ancestors is ubiquitous in all cultures, in one form or another. What evolutionary purpose it may play is not so clear.

Also difficult to explain from an evolutionary point of view are modern trends. For example, people in rich countries (from Japan to Western Europe) tend to have fewer children than people in poor countries. The usual explanation is that poor parents need to make many children in order to maximize the chances that at least one survives. But that does not explain why rich people would not make as many: if the goal is to maximize the probabilities of the survival of one's genes, wealthy people should make a child a year, since they can afford it. Humans are the only animals that have fewer babies when they are better fed.

Memes

The British biologist Richard Dawkins pointed out that there could be life beyond life when, in 1976, he introduced the cultural analogue of a gene (or, better, the analogue of genes for cultural transmission): the "meme". A meme is a unit of cultural evolution, just like a gene is a unit of biological evolution. A meme is an idea that replicates itself from mind to mind, such as a slogan or a refrain or a proverb. The US psychologist Donald Campbell had already argued that what we call "thinking" is really about ideas that are generated and then selected in a manner analogous to what happens to species ("Blind variation and selective retention in creative thought as in other knowledge processes", 1960).

Darwinian evolution is a process in three steps: replication (copying information), variation (making mistakes) and selection (pruning mistakes). Dawkins believes that what causes evolution is the genes: the "thing" that is copied, with mistakes, and then selected is genetic information.

From this premise, it was not difficult to conclude that "ideas" satisfy the same process of evolution: an idea is a pattern of information that is copied (from one mind to another), with some mistakes (each mind gets its own version) and then selected (some minds will be better than others at surviving, thanks to that idea, and at copying their version to other minds).

Both genes and memes are replicators.

The French biologist Jacques Monod had already noted the similarity between the spreading of genes and the spreading of ideas.

The human species is unique in that it relies on cultural transmission of information, and such a process is carried out by memes, the units of cultural evolution.

A meme is an idea that reproduces itself like a parasite. When a meme enters a mind, it parasitically alters the mind's process so that a new goal of the mind is to propagate the meme to other minds. The difference between a virus and a meme is that a meme is not an aggregate of DNA molecules but a structured piece of information, a piece of information that forces the mind to help reproduce it. Memes are behind the spreading of cults, fashions, ideologies and songs. The main difference between memes and viruses is really the stuff they are made of: viruses are made of DNA, whereas memes are made of information patterns.

Actually, at this point, you, the reader of this book, might just have been infected by a meme, the meme of memes. You will tell a friend what a meme is, and he will tell someone else, and so forth. The concept of what a meme is will spread from friend to friend, and, if it is interesting to the people who absorb it, possibly evolve.

Memes can be considered forms of life, or at least they behave like forms of life. Memes behave in a very similar way to genes. Genes are biological replicators, memes are nonbiological replicators.

Memes form an ecosphere of ideas.

The mind can be viewed as a machine for copying memes, just like the body is a machine for copying genes.

The US philosopher Daniel Dennett goes as far as to suggest that memes may have created the mind, not the other way around; that the mind was created by culture, not the other way around. Consciousness may simply be a collection of memes that is implemented in the brain as a sort of software in a machine that evolved in nature. Meaning itself would then be an emergent product of the meaningless algorithm that carries out evolution.

What makes us a superior species is not anatomy (which is roughly identical to the anatomy of a chimpanzee) but an odd plasticity of the brain that makes is more vulnerable to memes than any other species' brain. Humans are so "smart" because the human brain can be easily invaded by memes.

Just like genes use bodies as vehicles to spread and survive, so memes use minds as vehicles to spread and survive. Just like it is genes that drive evolution, it is memes that drive thought.

Memes are nonbiological replicators and they obey laws similar to the ones obeyed by biological replicators. There is one difference, though: that memes exhibit Lamarckian inheritance. Acquired characteristics can be passed on to future generations (ideas can be taught to children). This explains why ideas can evolve so much faster: Lamarckian evolutionism is "directed" and therefore much faster than Darwinian evolutionism.

A variation on memes was proposed by the US anthropologist Gregory Bateson. The mind is an aggregate of ideas. Ideas populate the mind and continuously evolve. Ideas evolve in a Darwinian fashion, the most useful ones surviving while useless ones decay and die away. Thus Bateson views the mind as the theater of a natural selection and evolution of ideas. Our conscious life "is" that evolutionary process.

Dawkins also grounded his theory of memes in Cairns-Smith's theory that self-replication originated in clay-crystals. According to Cairns-Smith, RNA came before DNA, and it was originally simply a passive element, but then eventually took over the self-replicating chore because it was more efficient than crystals; and then evolved into today's DNA. Dawkins simply abstracted this idea. A replicator needs to build a survival machine for itself (the body). Indirectly, that survival machine opens new possibilities for self-replications, just like, accidentally, crystals invented a more efficient kind of self-replicator, RNA. Dawkins thinks that we are now on the verge of a new genetic takeover, whereby "memes" will replace "genes" as the main self-replicating device. Memes are patterns of information. Their survival machines are brains, or, better, minds (Dawkins does not rule out that, for example, computers could act as survival machines for memes).

One can view the interaction between genes and memes at work in every family. No matter how much parents and children disagree on daily issues, parents program children to inherit the customs, traditions and beliefs of the "tribe", and children become enthusiastic carriers of those memes. Christians are proud of being (having been raised) Christian, and Muslims are proud of being (having been raised) Muslims, and Italians of being

Italian and Mexicans of being Mexican. It looks like genes create organisms whose function (or at least one of their functions) is to defend and propagate the culture of their parents. In a sense, genes program the body to reproduce the mental states of the parents and to pass them on to their own children. It is the interaction between genes and memes that accounts for the history of human civilization.

More Than One Evolution

If this theory has to be taken literally, then memes must be competing for survival, just like genes. This means that a meme does not necessarily serve the goals of a gene. A meme must be as selfish as a gene is.

The British psychologist Susan Blackmore believes that genes and memes actually "co-evolve".

She subscribes to the notion that each mind is but a meme machine ("a memeplex running on the physical machinery of a human brain"). Human action is the product of interactions between genes, memes and their environment.

A "memeplex" is a group of memes that band together for some mutual advantage. They assimilate memes that are compatible with them and reject memes that are incompatible. This way the memeplex as a whole becomes stronger and stronger and each participating meme benefits. Religions and ideologies are memeplexes.

Blackmore paints a picture of minds invaded by memes all the time, that function only as processors of memes.

Our minds are triggered by memes. We can never stop thinking. We do not think, we are thought by the memes that invade us.

As Dawkins has shown, memes are replicators. Blackmore formalizes this view in an extension of modern synthesis: Darwinian thinking must be applied to two replicators, not just one, and the result is meme-gene co-evolution. There are phenomena that cannot be explained by genetic motivation alone (e.g., language, which does not seem to provide any genetic advantage) but are easily explained by memetic motivation. Humans evolved along not one but two axes: the genetic one and the memetic one.

Language spreads memes, therefore language evolved to better spread memes. Language does not represent an evolutionary advantage for genes, but for memes.

Both genes and memes are replicators with equal status. The evolution of the human race is driven by evolution of two replicators.

"Body design" is achieved through competition between genes, i.e. genes compete to be passed to another body and in the process the body is shaped.

"Mind design" is achieved through competition between memes: memes compete to be passed to another mind and in the process the mind is created.

Both genetic and memetic factors are needed to explain what we are. In general, genes give us some skills, and then memes determine how we use them. For example, one can be genetically gifted as a writer, but then it is memes that determine what she will write.

Cultural Group Selection

If genes evolve and memes evolve, one has to wonder what is the connection between the two evolutions. The Russian geneticist Theodosius Dobzhansky had already noted that human evolution cannot be understood as a purely biological process.

The US anthropologists Robert Boyd and Peter Richerson point out that the single most important difference between humans and other species is the highly-developed ability, and the consequent dramatic consequence, of transmitting culture so that it affects other individuals. Cultural evolution happens at a much faster pace than genetic evolution: it does not require millions of years to unfold. Being a symbolic species, humans are able to create the abstraction of the "social group". A social group is, de facto, a cooperating group, regardless of whether the members of the group are kin or even know each other at all. It is this peculiarity of the human race that allowed the scale of societies to increase from the original tribal nucleus to full-fledged civilizations.

The US anthropologist William Durham argues that humans possess at least two information systems, one genetic and one cultural (and one has to wonder whether there are more). Culture is learned information that is conveyed socially and symbolic in nature (as in "language"), that has evolved over time, and that comes to constitute a system of knowledge. Human behavior is determined by two main information systems, one genetic and one cultural, that both spread over time and space. In other words, both the genetic repertory (the genotype) and culture are information systems that instruct phenotypes. The genotype and the culture stand in a symmetrical relationship to the environment, in that the environment selects "what" gets transmitted (inherited) in space and time. So genotype and culture stand in a symmetrical relationship with both the phenotype (each of them "instructs" it, although in different ways) and with the environment (each of them is "selected", although in different ways). Durham believes that two main forces are responsible for spreading culture across space and time: "selection by choice" and "selection by imposition". Choice is unique to culture because parents cannot choose which genes to pass on to their children, nor can children choose which genes to accept, whereas cultural parents can choose what to pass on and cultural children can choose what to accept. Imposition arises from the intervention of an intermediary that does not exist in the case of genetic transmission: sociopolitical constraints. More importantly, cultural evolution exhibits another unique property: self-selection. The cultural system can influence the direction and rate of its own evolution: memes influence human decisions that influence memes. The cultural fitness of an "allomeme" (a variant of a meme) indirectly depends on the meme itself.

Theory of Mind

One of the most impressive features of the human brain is its ability to understand other brains, i.e. the ability to construct a "theory of mind" about other people's intentions and feelings. By observing someone's face and behavior, human brains can infer that person's invisible "state of mind", an expression originally introduced by US psychologists David Premack and Guy Woodruff ("Does the chimpanzee have a theory of mind?", 1978). In fact, we can't help it: we building theories of mind whenever we look at somebody's face, i.e. we try to infer the mental state of every person we meet simply from looking at that person (typically, the eyes). The most important part of a portrait is the eyes: it is the first part of the portrait that we focus on, whether we like it or not. The eyes are the main clue about the mental state of the person.

We don't just build a theory of our own self, but also a theory of other people's selves. Children cry when other children cry, a sign that they identify with their pain. We feel sorry for other people's misfortunes (although not always happy for other people's luck). Premack also discovered that children are prewired with the distinction between "minds" and non-minds, i.e. between sentient beings and inanimate matter. Children treat differently objects that move by themselves and objects that move only when someone moves them. Children tend to see a "motive" behind self-propelled objects. It is a built-in ability to guess the state of mind of another being. Face perception might be the most developed visual skill in humans.

The Origin of Empathy

Laughter is contagious: if you look at a group of people who are laughing hysterically, you tend to laugh too, even if you cannot hear a single word of their conversation. Panic is contagious too: if you see a crowd of scared people, you get scared too.

There has long been consensus that each mind contains a theory of other minds: my mind constructs a theory of what is going on in your mind.

This "theory of mind" is crucial in interacting with others. It is also crucial for understanding your motives, which, in turn, is a crucial process for the many games of deceit that we play all the time (for example, bargaining or romancing a girl).

However, this "theory of mind" is more than just a summary or interpretation of the state of one's mind: it is a physical representation inside my brain of the neural state of somebody else's brain. My brain contains a "mirror image" of part of your brain. This explains the empathy: I feel your joy or your pain because my brian physically "duplicates" that brainstate and therefore makes me feel what you are feeling. Imitation and empathy are therefore built-in features of the brain.

Culture probably fine-tunes a mechanism that is there from birth. When parents "teach" their children, they craft the "mirror neurons" of their children, the same way that they teach their children to speak a language by fine-tuning the innate language skills of their brain. Children born with similar brains (like siblings) and raised by the same parents are inevitably likely to identify with each other's states of mind. Similarly, when a teacher teaches the values of society to a class, the mirror neurons of the pupils are being shaped to create a uniform brain state towards those values.

"Mirror neurons", discovered by the Italian neurologist Giacomo Rizzolatti ("Action Recognition in the Premotor Cortex", 1996), might be affected by any kind of communication with other people. As you live in your environment, the environment trains your mirror neurons to reflect its values.

Mirror neurons are a possible explanation for the power of peer pressure and of fads. Mirror neurons explain altruism: the single most powerful motivation to help someone in trouble is that I can feel their pain and the only way to stop feeling it is to help them get out of trouble. Then I will feel their joy.

Mirror neurons might be the single most important feature for creating societies and civilizations.

Mirror neurons do not reset overnight. They are "remembered" just like anything else. This means that my mind contains (part of) the minds of all the people encountered in my lifetime. My mind contains your mind, in fact it contains all the minds with which I have interacted, although only as much as was "revealed" during that interaction. My "self" contains more than me.

Cooperation is rather primitive among other mammals and birds. The reason could be that only human brains evolved such sophisticated mirror neurons.

Today mirror neurons may in fact be working beyond their original mission, causing us to sympathyze with and help people even in circumstances that may harm us; something that was certainly not the original reason for mirror neurons to evolve.

The US psychologist Paul Bloom thinks that morality is hard-wired in the human brain. Empathy is born with life: when babies hear crying, they start to cry themselves. They "feel" the pain of other babies. We feel the pain of other people. It is just natural that we desire for that pain to go away because it affects us, not only them. Morality and altruism are not learned. Humans are endowed at birth with a "moral sense".

The Origin Of Evil
The US psychologist Howard Bloom believes that Darwinian evolution can explain the history of humankind. However, he sees evolution as operating on a hierarchy of systems.

In Bloom's Darwinian world, it is societies, and not only individuals, that compete for survival. Societies are capable of amazing feats that seem due to an invisible mind, but are simply due to a "neural network" of individuals. Societies are super-organisms that develop their own mental lives. Memes are sort of the "thoughts" of these super-organisms. After all, isn't the human body just a society of cells that could live independently but have decided to live together? In all aspects of civilization he sees the working of a superorganism, a network of individuals, a network that takes on a life of its own, driven by Darwinian competition.

Just like organisms, societies too compete against each other. Evil is the inevitable consequence of that kind of global competition among societies. He notices that every society regards outsiders as not really human and justifies hatred as righteousness. Why does history repeat itself? Because the same evolutionary principles are at work, over and over again, in each and every society.

The Neurological Alternative to Sociobiology
The French archeologist Jacques Cauvin introduced the expression "symbolic revolution" to refer to the sudden change in art and society that took place in the neolithic, notably the transition from hunting and gathering to agriculture and domestication of animals, with the consequent transition from nomadic life to settled life. The traditional explanation was that whatever caused the change, it came from the environment and then it "migrated" into people's minds, causing a new way of symbolic thinking in religion and politics. Cauvin argued that, on the contrary, agriculture and domestication of animals were a by-product of a change in mental life: first the mind underwent the symbolic revolution, which also involved a new cosmology, and then this new mind conceived of agriculture and domestication. The change in mental life must have originated from a physical modification of the brain, from a mutation of sorts.

Along those lines South African archeologists David Lewis-Williams and David Pearce claimed that human neurology inevitably leads to a tripartite nature of religion, which yields a three-realm cosmology that they believed to be widespread in ancient civilizations.

An extreme interpretation of this argument would be that humans created civilization not to adapt to the environment and not because it was useful to survival but simply because a mutation in their brains made them do those things, just like human brains make humans see and hear. Everything that happened afterwards was just a consequence of a new brain that was thinking differently.

Further Reading

Axelrod, Robert: THE EVOLUTION OF COOPERATION (Basic Books, 1984)

Bateson, Gregory: STEPS TO AN ECOLOGY OF MIND (Chandler, 1972)

Blackmore, Susan: THE MEME MACHINE (Oxford Univ Press, 1999)

Bloom, Howard: THE LUCIFER PRINCIPLE (Norton, 1995)

Bloom, Paul: Descartes' Baby (Basic, 2004)

Boyd, Robert and Richerson, Peter: CULTURE AND THE EVOLUTIONARY PROCESS (Univ of Chicago Press, 1985)

Boyd, Robert and Richerson, Peter: THE ORIGIN AND EVOLUTION OF CULTURES (Oxford Univ Press, 2005)

Brandon, Robert: GENES ORGANISMS POPULATION (MIT Press, 1984)

Brandon, Robert: ADAPTATION AND ENVIRONMENT (Princeton Univ Press, 1990)

Buss, David: THE EVOLUTION OF DESIRE (Basic, 1994)

Capra, Fritjof: THE WEB OF LIFE (Anchor Books, 1996)

Cauvin, Jacques: NAISSANCE DES DIVINITES, NAISSANCE DE L'AGRICULTURE (1994)

Cavalli-Sforza, Luigi & Feldman, Marcus: CULTURAL TRANSMISSION AND EVOLUTION (Princeton Univ Press, 1981)

Corning, Peter: NATURE'S MAGIC (Cambridge Univ Press, 2003)

Cosmides, Leda & Tooby, John: THE ADAPTED MIND (Oxford Univ Press, 1992)

Cronin, Helena: THE ANT AND THE PEACOCK (Cambridge University Press,1992)

Darwin, Charles: THE DESCENT OF MAN, AND SELECTION IN RELATION TO SEX (1871)

Dawkins, Richard: THE SELFISH GENE (Oxford Univ Press, 1976)

Dawkins, Richard: THE BLIND WATCHMAKER (Norton, 1987)

Dawkins, Richard: RIVER OUT OF EDEN (Basic, 1995)

Dawkins, Richard: CLIMBING MOUNT IMPROBABLE (Norton, 1996)

De Waal, Frans: GOOD NATURED : THE ORIGINS OF RIGHT AND WRONG IN HUMANS AND OTHER ANIMALS (Harvard University Press, 1996)

De Waal, Frans: THE AGE OF EMPATHY (Broadway Books, 2009)

Dennett, Daniel: DARWIN'S DANGEROUS IDEA (Simon & Schuster, 1995)
Dobzhansky, Theodosius: MANKIND EVOLVING (1962)
Durham, William: COEVOLUTION (Stanford Univ Press, 1991)
Frisch, Karl von: THE DANCE LANGUAGE AND ORIENTATION OF BEES (Harvard University Press, 1967)
Goertzel, Ben: THE EVOLVING MIND (Gordon & Breach, 1993)
Gordon, Deborah: ANTS AT WORK (Free Press, 1999)
Gould, Stephen Jay: THE STRUCTURES OF EVOLUTIONARY THEORY (Harvard Univ Press, 2002)
Hamilton, William Donald: NARROW ROADS OF GENE LAND (W.H. Freeman, 1996)
Hamilton, Terrell: PROCESS AND PATTERN IN EVOLUTION (MacMillan, 1967)
Hauser, Marc: MORAL MINDS (Ecco, 2006)
Hinde, Robert: WHY GOOD IS GOOD (Routledge, 2002)
Hölldobler, Bert and Wilson, Edward-Osborne: THE SUPERORGANISM (2008)
Johnson, Cecil: MIGRATION AND DISPERSAL OF INSECTS BY FLIGHT (1969)
Jolly, Alison: LUCY's LEGACY (Harvard University Press, 1998)
Keller, Laurent: THE LIVES OF ANTS (Oxford Univ Press, 2009)
Kinji, Imanishi: THE WORLD OF LIVING THINGS (1941)
Krebs, Dennis: THE ORIGINS OF MORALITY (Oxford Univ Press, 2011)
Kropotkin, Petr: MUTUAL AID (1902)
Levi-Strauss, Claude: LE CRU ET LE CUIT (1964)
Lewis-Williams, David & Pearce, David: INSIDE THE NEOLITHIC MIND (Thames & Hudson, 2005)
Lewontin, Richard: THE GENETIC BASIS OF EVOLUTIONARY CHANGE (Columbia University Press, 1974)
Lovelock, James: GAIA (Oxford University Press, 1979)
Margolis, Howard: SELFISHNESS, ALTRUISM, AND RATIONALITY (Univ of Chicago Press, 1984)
Margulis, Lynn: WHAT IS LIFE? (Simon & Schuster, 1995)
Maynard-Smith, John: THEORY OF EVOLUTION (Cambridge University Press, 1975)
Maynard Smith, John: EVOLUTION AND THE THEORY OF GAMES (1982)
Mayr, Ernst: TOWARDS A NEW PHILOSOPHY OF BIOLOGY (Harvard Univ Press, 1988)
Maynard-Smith, John & Szathmary, Eors: THE MAJOR TRANSITIONS IN EVOLUTION (W. H. Freeman, 1995)
McCullough, Michael: BEYOND REVENGE (2008)

Merezhkovsky, Konstantin: THEORY OF TWO PLASMS AS THE BASIS OF SYMBIOGENESIS (1909)
Miller, Geoffrey: THE MATING MIND (Doubleday, 2000)
Murchie, Guy: SEVEN MYSTERIES OF LIFE (Houghton Mifflin, 1978)
Nesse, Randolph & Williams, George: WHY WE GET SICK (Times Books, 1994)
Nowak, Martin: SUPERCOOPERATORS (Free Press, 2011)
Oparin, Alexander: THE ORIGIN OF LIFE (1924)
Premack, David: Original Intelligence: The Architecture of the Human Mind (McGraw-Hill, 2002)
Ridley, Matt: EVOLUTION (Blackwell, 1993)
Ridley, Matt: THE RED QUEEN (MacMillan, 1994)
Ridley, Matt: THE ORIGINS OF VIRTUE (Viking, 1997)
Ridley, Mark: THE COOPERATIVE GENE (Free Press, 2001)
Rifkin, Jeremy: THE EMPATHIC CIVILIZATION (Tarcher, 2009)
Shermer, Michael: THE SCIENCE OF GOOD AND EVIL (New York Times, 2005)
Shlain, Leonard: SEX, TIME, AND POWER (Penguin, 2003)
Sigmund, Karl: GAMES OF LIFE (Oxford University Press, 1994)
Sober, Elliot and Wilson, David Sloan: UNTO OTHERS (Harvard Univ Press, 1998)
Trivers, Roberts: SOCIAL EVOLUTION (Benjamin/Cummings, 1985)
Villarreal, Luis: VIRUSES AND THE EVOLUTION OF LIFE (ASM Press, 2004)
Von Frisch, Karl: THE DANCING BEES (1953)
Wallin, Ivan: SYMBIOTICISM AND THE ORIGIN OF SPECIES (1927)
Wheeler, William-Morton: SOCIAL LIFE AMONG THE INSECTS (1923)
Williams, George: ADAPTATION AND NATURAL SELECTION (Princeton University Press, 1966)
Wilson, Edward Osborne: SOCIOBIOLOGY (Belknap, 1975)
Wilson, Edward Osborne: CONSILIENCE (Knopf, 1998)
Holldobler, Bert & Wilson, Edward: THE SUPERORGANISM (Norton, 2008)
Wilson, Edward Osborne: THE DIVERSITY OF LIFE (Harvard University Press, 1992)
Wilson, Edward & Lumsden Charles: GENES, MIND AND CULTURE (Harvard Univ Press, 1981)
Wrangham, Richard: Catching Fire - How Cooking Made Us Human (Basic, 2009)
Wright, Robert: THE MORAL ANIMAL (Random House, 1994)

Wright, Robert: NON-ZERO (Pantheon, 2000)Wynne-Edwards, Vero-Copner: ANIMAL DISPERSION IN RELATION TO SOCIAL BEHAVIOUR (Oliver & Boyd, 1962)

Wynne-Edwards, Vero-Copner: EVOLUTION THROUGH GROUP SELECTION (Blackwell, 1986)

Alphabetical Index of Names

Abraham, Ralph : 93
Arrhenius, Svante : 46
Avery, Oswald : 32
Axelrod, Robert : 107
Baldwin, James : 60
Barwise, Jon : 8, 9
Bateman, Angus : 129
Bates, Elizabeth : 35
Bateson, Gregory : 19, 81, 138
Bateson, William : 31, 40
Behe, Michael : 65, 109
Blackmore, Susan : 139
Bloom, Howard : 143
Bloom, Paul : 142
Bogdan, Radu : 9
Boltzmann, Ludwig : 31, 75
Boyd, Robert : 140
Breitenberg, Valentino : 12, 34
Brenner, Michael : 54
Brenner, Sydney : 32
Brock, Thomas : 48
Brooks, Daniel : 88
Brooks, Rodney : 11, 13
Butler, Samuel : 70, 94, 114
Cairns, John : 68
CairnsSmith, Graham : 47
Calvin, Melvin : 45
Campbell, Donald : 137
Capra, Fritjof : 118
Carlson, Richard : 14
Cauvin, Jacques : 143
CavalliSforza, Luigi : 38
Cech, Thomas : 49
Clark, Andy : 12
Conrad, Michael : 45
Corning, Peter : 119, 120
Cosmides, Leda : 129
Crick, Francis : 32, 64
Crooks, Gavin : 54
Cziko, Gary : 33

Dano, Sune : 115
Darwin, Charles : 25, 27, 28, 31, 68, 80, 104, 109, 123, 130
Davies, Paul : 48, 93
Dawkins, Richard : 19, 25, 48, 66, 123, 125, 136, 137, 138
DeBeer, Gavin : 69
DeDuve, Christian : 54
Delsemme, Armand : 46
Dennett, Daniel : 8, 137
Deutsch, David : 43
DeVries, Hugo : 31
DeWaal, Frans : 107
Dobzhansky, Theodosius : 32, 140
Dretske, Fred : 8
Driesch, Hans : 93
Durham, William : 140
Dyson, Freeman : 50, 53, 82, 98
Edelman, Gerald : 91
Eigen, Manfred : 44, 50, 53
Elsasser, Walter : 93
England, Jeremy : 54
Evans, Denis : 54
Fisher, Ronald : 31, 32, 56, 131
Forterre, Patrick : 50
Fox, Ronald : 82
Frautschi, Steven : 99
Frege, Gottlob : 8
Fuller, Buckminster : 96
Gallistel, Randy : 10
Galton, Francis : 64
Ganti, Tibor : 82
Gerhart, John : 69
Gibson, James : 5, 7
Gilbert, Walter : 49
Goertzel, Ben : 110
Goldschmidt, Richard : 64
Goodwin, Brian : 91
Gordon, Deborah : 115

Index

Gould, Stephen : 34, 63, 64, 109
Grammer, Karl : 132
Haldane, John : 104
Hamilton, Terrell : 122
Hamilton, William : 104, 105, 106, 127, 129
Hughes, Howard : 20
Hull, David : 124
Hutchinson, Evelyn : 80
Huxley, Julian : 31
Ingber, Donald : 96
Islam, Jamal : 98
Jacob, Francois : 32
James, William : 33
Jansch, Erich : 76
Jarzynski, Chris : 54
Jeon, Kwang : 114
Johanssen, Wilhelm : 31
Johnson, Cecil : 115
Johnson, Lionel : 82
Johnson, Mark : 19
Jolly, Alison : 108
Jones, Steven : 37
Joyce, Gerald : 49
Kant, Immanuel : 19
Kauffman, Stuart : 45, 68, 91
Kestin, Joseph : 78
Khorana, Har : 32
Kinji, Imanishi : 106
Kirschner, Marc : 69
Kropotkin, Petr : 106
Kuppers, Bernd : 52, 81, 122
Lamarck, JeanBaptiste : 59
Lande, Russell : 132
Lane, Nick : 48
Langton, Christopher : 82
Laplace, Pierre : 52
Layzer, David : 85, 88
LeviStrauss, Claude : 132
Lewis, David : 143
LewisWilliams, David : 143
Lewontin, Richard : 20, 43, 122
Lieberman, Philip : 62
Lipmann, Fritz : 38, 77, 83

Lorenz, Konrad : 33, 34
Lotka, Alfred : 75, 77, 78, 82, 88
Lovelock, James : 18, 118
Luria, Alexander : 87
MacWhinney, Brian : 35
Marcus, Philip : 54
Margalef, Ramon : 80, 81
Margulis, Lynn : 17, 50, 111, 114, 118, 121
Marr, David : 6
Maturana, Humberto : 14, 110
Maxwell, James : 31
MaynardSmith, John : 56, 58, 107, 118
Mayr, Ernst : 61, 63, 123
McCulloch, Warren : 12
Mendel, Gregor : 30, 42
Merezhkovsky, Konstantin : 111
MerleauPonty, Maurice : 18
Mikulecky, Don : 79
Miller, Geoffrey : 130
Miller, Stanley : 44, 49
Millikan, Ruth : 19
Monod, Jacques : 32, 51, 53, 120, 137
Morowitz, Harold : 53, 62, 77
Muller, Anthonie : 46
Murchie, Guy : 116, 121
Neisser, Ulric : 7
Newton, Isaac : 31
Nirenberg, Marshall : 32
Nowak, Martin : 107
Odum, Eugene : 77, 80
Odum, Howard : 77
Oparin, Alexander : 44
Oro, John : 46
Parter, Merav : 69
Pasteur, Louis : 49
Pattee, Howard : 66
Pearce, David : 143
Perry, John : 8
Pitts, Walter : 12
Plotkin, Henry : 8, 101
Powers, William : 34

Premack, David : 141
Prigogine, Ilya : 75, 78
Rapaport, Anatol : 107
Rebek, Julius : 48
Richerson, Peter : 140
Ridley, Mark : 125
Ridley, Matt : 109, 126
Rizzolatti, Giacomo : 142
Scaruffi, Piero : 1, 2
Schmalhausen, Ivan : 87
Schneider, Eric : 78
Schroedinger, Erwin : 61, 75, 78
Sheldrake, Rupert : 93
Shlain, Leonard : 133, 135
Sigmund, Karl : 107
Snelson, Kenneth : 96
Speman, Hans : 93
Spencer, Herbert : 116
Szathmary, Eors : 56, 118
SzentGyorgyi, Albert : 75
Szyf, Moshe : 40
Tarnita, Corina : 108
Teilhard, Pierre : 117, 118
Thom, Rene : 92
Thompson, DArcy : 89
Thornhill, Randy : 132
Tipler, Frank : 99
Tolman, Edward : 7
Trivers, Robert : 105, 106, 129
Turing, Alan : 89, 91
Varela, Francisco : 18

Vernadsky, Vladimir : 17, 118
Villarreal, Luis : 117
VonBertalanffy, Ludwig : 74
VonFrisch, Karl : 115
VonHelmholtz, Hermann : 5
VonUexkull, Jakob : 5
Waddington, Conrad : 61, 92, 93
Waechtershauser, Gunter : 44
Wallace, Alfred : 123
Watson, James : 32
Watson, John : 40
Weismann, August : 89, 108
Weiss, Paul : 10, 93
Weizsacker, Carl : 81
Wheeler, John : 116
Wheeler, William : 116
Wicken, Jeffrey : 48, 76, 123
Wiley, Edward : 88
Williams, George : 105, 129, 130
Williams, Richard : 123
Wilson, Allan : 38
Wilson, David : 106, 122
Wilson, EdwardOsborne : 62, 127
Wimsatt, William : 123
Woese, Carl : 48, 57, 111
Woltereck, Richard : 74
Woodruff, Guy : 141
Wrangham, Richard : 132
Wright, Robert : 120, 130
Wright, Sewall : 31

www.ingramcontent.com/pod-product-compliance
Lightning Source LLC
Chambersburg PA
CBHW051708170526
45167CB00002B/579